Bound to the Sun

The Story of Planets, Moons, and Comets

Bound to the Sun
The Story of Planets, Moons, and Comets

Rudolf Kippenhahn

Translated by Storm Dunlop

W. H. Freeman and Company
New York

Library of Congress Cataloging-in-Publication Data

Kippenhahn, Rudolf, 1926–
    [Unheimliche Welten. English]
    Bound to the sun: the story of planets, moons, and comets/
Rudolf Kippenhahn; translated by Storm Dunlop.
        p.   cm.
    Translation of: Unheimliche Welten.
    Includes index.
    ISBN 0-7167-2124-4
    1. Astronomy.   2. Solar system.   I. Title.
QB43.2.K5413   1990
523.2 — dc20   90-36174
                                                      CIP

Title of the original German edition: *Unheimliche
Welten: Planeten, Monde und Kometen.* Copyright © 1987
Deutsche Verlags-Anstalt GMBH, Stuttgart.

Printed in the United States of America

1 2 3 4 5 6 7 8 9 0   VB   9 9 8 7 6 5 4 3 2 1 0

Contents

Preface

This book arose from a course of lectures that I gave in the summer term of 1985 for students from all the various faculties at the Ludwig-Maximilian University in Munich. In these lectures, I tried to describe our knowledge of the bodies that are gravitationally bound to the sun and which orbit in our immediate vicinity.

Within my family and circle of friends, we have often discussed the question of whether or not the fictional Herr Meyer should appear in my books. I think that in the present work he fulfils his purpose. I use him to help my readers experience the world of the planets. I have not the slightest intention, however, of turning the Herr Meyer stories into a form of science fiction.

Many colleagues and friends helped me to eliminate errors from my original text. The late Richard-Heinrich Giese and my colleagues, Reinhold

Häfner, Gerhard Haerendel, Wolfgang Hillebrandt, Peter Kafka, Gregor Morfill, Felix Schmiedler, Jürgen Teichmann, Hans-Heinrich Voigt, and Richard Wielebinski, have read individual sections of the book and suggested changes. Wolfgang Duschl and Rhea Lüst read through every chapter. The mathematician Hans-Ludwig de Vries, who is simultaneously my friend and severest critic, went through the whole manuscript with me in a fortnight of concentrated effort.

Horst Uwe Keller and Rüdiger Reinhard enabled me to be at Darmstadt during the critical period of the "Night of the Comet." Cornelie Rickl helped me with my correspondence and undertook what was often hours of work printing out my diskettes. The designer Jutta Winter and the "crew" of the Deutsche Verlags-Anstalt made a very considerable contribution to the success of the book.

The manuscript would never have been finished if my wife had not been, on the one hand, critical, and on the other, constantly encouraging.

The hospitality of the Remeis Observatory at Bamberg was a great help, allowing me to bury myself in their library and work on the manuscript.

The staff of W. H. Freeman and Company produced the English-language edition, which was not an easy task since the publisher was on one continent and the author and the translator on another.

In listing all the help that was so kindly given to me, I really wonder how much there was left for me to do. I am very grateful to everyone concerned.

Rudolf Kippenhahn
Munich, April 1990

Bound to the Sun

The Story of Planets, Moons, and Comets

Introduction

■

The year was 1504. Christopher Columbus' fourth journey to the New World had already lasted about two years when a storm left the crew stranded on the Jamaican coast. Worms had eaten the timbers of the half-wrecked ships, which the castaways dared not use to reach Hispaniola, the next island. Against their captain's orders, many crewmembers plundered the Indian villages and tried to escape in stolen canoes. Only a few faithful men remained with Columbus, knowing that at any moment the natives could attack their stockade.

Soon, even these men grew weary of listening to orders from their sick, tired leader, prematurely aged for his 53 years. It was to be Columbus' last voyage. Yet it was important to be prepared for the Indians, who realized long ago that the foreigners were falling out among themselves. The belief that the newcomers were semidivine had also disappeared, until by chance, a heavenly event came to their aid.

To navigate, Columbus had been using the astronomical tables written by the Franconian astronomer Johannes Müller (1436–1476), better known by his Latin name of *Regiomontanus*. By the time Columbus found himself in serious trouble in Jamaica, Regiomontanus had been dead for 28 years. However, his tables, which predicted the position of the celestial bodies for the years 1475 to 1506, had survived and were still used for navigation.

Regiomontanus' book also contained detailed information on eclipses, predicting a total eclipse of the moon on 29 February. Thus, on that day, Columbus invited the Indian chiefs to meet him in his stockade. His exact words were not recorded, but the translator probably said something to this effect: "The Christian god rewards the good and punishes the evil. He is angry with you for letting us, His messengers, starve and suffer. For this, He will punish you severely. Out of His great mercy, however, He will give you the chance to repent. As a sign of His wrath, He will take away the light of the moon tonight; and if you do not heed His warning, the moon will disappear from the sky forever." Not all the natives were as impressed by his threat as Columbus had hoped. A few appeared to be shocked, but others just laughed. Columbus himself was extremely worried, because he did not know if Regiomontanus' tables were reliable.

Fortunately, the predictions by the Franconian astronomer held true. Shortly after the full moon rose in the east, a shadow started to creep over it. The higher the moon rose, the clearer the shadow could be seen. Soon there was nothing left but a reddish disk glimmering in the sky. The Indians ceased their mockery. Their chiefs threw themselves on the ground and begged Columbus to ask his god to restore the light of the moon. Columbus agreed to see what could be done and retired to his hut. There he measured the duration of the eclipse with a half-hour sand-glass, then went back outside and announced that his god would be forgiving. And indeed, the moon did return to its normal brightness.

The next day no one had to go hungry. The natives outdid themselves trying to please the strangers, who had demonstrated that they could speak directly with a mighty god. Not long afterwards, the castaways were finally rescued by a Spanish ship.

Since then, the idea of being saved by an eclipse has been a popular theme in adventure novels. Many brave heroes languishing in the hands of primitive enemies have received help from the sky. For example, Mark Twain used a similar event, adapted as a solar eclipse, in *A Connecticut Yankee in King Arthur's Court*. It is amazing that of the few eclipses that occur in any given year, one should happen at just the right time.

But, returning to the story of Columbus, what made that event so remarkable was not the element of coincidence, but the fact that Regiomontanus' predictions were accurate at all. Neither Regiomontanus nor Columbus believed that the earth and the other planets orbited the sun. The accepted view was that the sun, the moon, and the planets orbited the earth in a complicated pattern. Not until 67 years after Regiomontanus' death did Copernicus (1473–1543) prove otherwise. It seems almost miraculous that without knowing the true nature of their orbits, Regiomontanus could correctly predict the movements of the heavenly bodies and Columbus could depend on these tables to save his crew.

Columbus did not know what we learn in school today: The planets move in almost circular orbits around the sun. They are not luminous in themselves and can therefore only be seen by the sunlight that is reflected from their surfaces. The innermost orbit is that of Mercury, which orbits the sun in 88 earth-days. The outermost planet was only discovered in 1930 by a young farm boy from Kansas. This new planet requires 248 earth-years to complete one orbit. Since its discovery, it has only covered about one quarter of its orbit. It is so far out in space that from its distance, the sun appears merely as a bright star in the sky. Between the innermost and the outermost planets, there are seven large bodies and innumerable smaller ones.

In order from the sun, the planets are called Mercury, Venus, Earth, Mars, Jupiter, Saturn, Uranus, Neptune, and Pluto. Those who find it difficult to remember their order may be helped by the mnemonic: "Men Very Easily Make Jugs Serve Useful Nocturnal Purposes." In addition, there are numerous small bodies, or miniplanets, not much larger than a few kilometers in diameter, and comets that do not follow almost circular orbits like the planets, but which arrive from far out in space, cut across the region of the planets, and then vanish again into the depths of space.

Even the larger planets are small compared with the distances between them. Like motes of dust, they move in an unimaginably large region of space. They are also small in comparison with the sun, which rules them all with its gravity. The Göttingen astronomer Hans-Heinrich Voigt once gave a striking comparison in a lecture for beginners: Assume that we want to make a reduced-scale model of the solar system and that 1000 kilometers correspond to 1 millimeter in our model. Then the sun would be a sphere 1.4 meters across. Mercury would be the size of a pea at a distance of 60 meters. Venus and Earth would be hazelnuts at distances of 110 meters and 150 meters, respectively. Mars would also be the size of a pea, orbiting 230 meters away from the sun. Jupiter and Saturn would be cabbages, 14 and 12 centimeters across, respectively, and 800 meters and 1400 meters away. Uranus and Neptune would be approximately equal-sized mandarin oranges at distances of 3 and 4.5 kilometers, respectively. Pluto, another pea, would have an average distance of 6 kilometers from the sun. That 6 kilometers would encompass our solar system. On this scale, the nearest star would lie at a distance of about 40,000 kilometers, and in our model, it would not even be possible to show it on the surface of the earth.

Today we find the arrangement of the planets, the nature of their motion, and their orbits perfectly natural, and we learned it in school. However, it took people thousands of years to reach this conclusion. Nevertheless, such knowledge is by no means universal among our fellow citizens. I suspect that in America and Europe more people know what sign of the zodiac they were born under than remember that the moon orbits the earth, and Mars the sun.

Our knowledge of the motions of celestial bodies dates from the sixteenth century. Since then, people have not only found bodies that move around the sun and whose existence had never previously been suspected, but also discovered that many planets have moons and that some have rings in which smaller lumps of ice and rock orbit the planets themselves. We have discovered the chemical composition of the atmospheres of planets and satellites. Instruments sent to the surfaces of Venus and Mars have radioed back information, including details of the temperature and pressure on Venus, and pictures of the desert surface of Mars. The thick layers of clouds around Venus have been pierced by radar, and the echoes of radio signals sent out from Earth have determined the length of the day on Mercury. Men have walked around on the moon; space probes have passed through Saturn's rings and observed erupting volcanoes on Jupiter's satellite Io. The *Giotto* space probe went within 600 kilometers of Halley's comet, before sandblasting by the cometary dust perforated its protective shield.

Space probes have transmitted wonderful color pictures of Jupiter, Saturn, Uranus, and Neptune into our very homes, which make the planets seem harmless enough places. However, the surfaces of these four gas giants are poisonous, foul-smelling worlds, where no humans would survive for a minute. The expanses of Mars are waterless deserts, and the landscape on Venus is a swelteringly hot hell, where even specially prepared instrument capsules, protected from the sulfuric-acid rain that falls from the sky, cease to work after a short period of time. The temperatures found on the icy wastes of the outer planets and their satellites make the polar regions of the earth seem positively cozy by comparison.

But thousands of years before mankind reached this stage, there were people who puzzled over the drama that was being enacted in the night sky before their very eyes.

ONE

The Heavenly Machine

As long as the earth remained stationary, all true astronomy was—and had to remain—static; however, as soon as the man appeared who caused the sun to stand still, astronomy began to move forward.

GEORGE CHRISTOPH LICHTENBERG (1742–1799)

■

Because their days were unbearably hot, the Chaldean herdsmen waited until nightfall to drive their herds out to pasture. To guide the way, these ancient herdsmen depended on the points of light scattered unevenly across the sky, giving them names that we no longer know or use. Night by night they watched the stars follow one another across the sky, slowly swept along by some unseen current. They also discovered that other stars did not follow the same regular pattern, but moved differently in a way that puzzled the stargazers. These early Chaldean herdsmen had no idea what they were watching. In fact, thousands of years passed before man really understood how the stars moved.

■ The Movements in the Sky

The beginnings of astronomy go back some 4000 years and are forgotten today. Undoubtedly, people had looked at the night sky even earlier and had been unable to understand what they saw on the black vault that slowly circled the earth, causing the stars to rise in the east and to set in the west night after night. The nightly performance was regularly interrupted when the sun, the brightest object in the sky, rose above the horizon in the east, driving away the darkness of the night. The sky then lost its glittering stars and became uniformly blue. The Dutch mathematician and historian of science, Bartel L. van der Waerden, wrote:

> Every race knows that the sun rises in the east and sets in the west, that it rises higher and shines longer in summer than in winter.
>
> Everyone calculates time in days, months, and years. Common experience teaches us that the moon is first visible as a thin crescent in the evening, that after about 14 days it shines throughout the night as the full moon, and that it disappears after a further 14 days. The daily rotation of the sphere of the fixed stars around the polestar is visible to everyone. It needs no very astute observer to see that Venus and Jupiter are brighter than other stars and that they do not always remain in the same constellation.

These observations were probably not hidden from prehistoric people, either. They must have felt that a mighty pageant was being played out before their eyes. Myths arose, expressing the coming and going of the stars in poetic form. Four thousand years ago, the Chinese tried to explain nature as the interplay of Yang, the heavenly element, and Yin, the earthly one, an explanation that can be found in an old Chinese book of divination. It was believed that something about the future could be learned from the stars. In those days, the stars appeared to be gods, or at the very least, mysterious entities, not a part of the everyday, real world.

Inside the lid of a 3000-year-old Egyptian sarcophagus, there are details of the times of the rising of Sirius. The true science of astronomy began as soon as people began to record observations of the stars and to compile lists of the times of their rising. Such lists are also known from Babylonia. What could be seen in those ancient times and what could be measured and recorded about the events that occurred?

▪ The Rhythm of the Day

Let us put ourselves in the place of those early people, who were familiar with the night sky. Nowadays we have telescopes at our disposal, and even with a simple pair of binoculars, we have a considerable advantage over them. Let us, therefore, lay aside telescopes and binoculars. Without these aids, we realize that those early people had an advantage over us, because they not only had a clearer sky, unpolluted by artificial light, but they also had more time to spare. It required patient observation over years and decades to learn the laws that governed the vast clockwork of the heavens.

So what did the Chaldean herdsman, who looked at the sky night after night, actually see? Let us, like him, patiently watch the sky.

When the sun sinks below the horizon in the west and night falls, the stars appear. They do not stand still, but move across the heavens from east to west throughout the night, just as the sun follows its path across the sky during the day. Stars that were high above the horizon at nightfall, as well as those seen in the western sky, follow the sun down below the western horizon. Their place is taken by other stars rising above the horizon in the east. The movement is so slow that we are unable to see it directly. It is easy to recognize, however, if we look at a particular constellation, say, two hours later. This movement does not affect the position of stars relative to one another. It is as if they were fixed to a starry sphere, and their movement came from the sphere rotating around us.

Not all stars rise and set. Two points in the sky remain in exactly the same place throughout the night. These are the *celestial poles,* around which the celestial sphere appears to rotate. One, the north pole, can be seen from the northern hemisphere of the earth. Or, more precisely, we can look at the point in the sky about which everything appears to rotate. There is nothing actually there that our eyes can use as a reference point. But at a distance of about twice the diameter of the full moon, there is a star that can be clearly seen with the naked eye: the North Star. During the night, the stars that we can see from the earth's northern hemisphere rotate around the north celestial pole. From the latitude of Washington, D.C., this one fixed point among the rotating stars appears to be about three and a half handwidths above the northern horizon. As they rotate around the pole, only those stars that are farther from it than this distance appear to rise and set. Although the stars closer to the pole, such as the North Star itself, do rotate about the pole, they always remain above the horizon, like the pole itself. This motion of the stars is clearly visible if we take a camera, direct it toward the north, and take a time exposure of a few hours, without changing the position of the camera. Such a picture is shown in Figure 1.

At dawn, the light of the rising sun interrupts the stars' perpetual round, but then the sun itself follows the same path across the sky from east to west. When it sets in the evening and night falls again, everything is just as it was the evening before. The stars that could be seen in the twilight the day before reappear in roughly the same places and follow the same paths as they rise and set. Everything still turns around the north celestial pole, and the stars that were too close to it to set yesterday remain above the horizon today.

▪ The Rhythm of the Year

If we make this observation regularly, we shall soon see that everything is a little bit more complicated than it first seems. The sky is not quite the same

Figure 1. *A photograph of the night sky, taken with a stationary camera from central Europe, with an exposure of 85 minutes. During this time, the stars moved across the sky and therefore appear drawn out into streaks. It is easy to see that all of the stars are moving in arcs, and that the central point of all the arcs is the northern celestial pole. The apparent motion of the stars is, of course, a result of the rotation of the earth, which carries the camera with it (Photo: Volkmar Voigtländer).*

night after night. After a few weeks, stars that initially had only just risen at twilight are found to be high up in the evening sky at dusk. Stars that were initially low in the western twilight and set shortly afterwards are no longer visible in the night sky during the weeks and months that follow. They will only reappear in the morning twilight about three months later. Stars that first rose just at dawn, only to be drowned out almost immediately by the light of the rising sun, after a few weeks will rise during the night. Week by week, they remain longer in the sky before they disappear in the daylight. If we observe more closely, we shall see that in comparison with the twilight, and thus with the sun, the stars rise and set about four minutes earlier each day. In the course of a year, the 365 × 4 minutes add up to exactly one day.

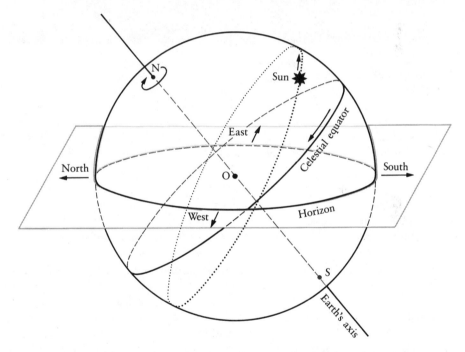

Figure 2. *To an observer (shown here at point O) at any northern latitude on the earth's surface, the sky appears as a hemisphere lying above the circle formed by the horizon. The sphere of the fixed stars rotates around its axis once a day. The two celestial poles (N and S) lie on this axis. The direction of rotation is shown by arrows on the celestial equator and at the north pole. This rotation causes the stars on the celestial sphere to rise at the eastern horizon and set in the west. The sun rises and sets with the stars. It does not, however, remain in the same place on the sphere, but moves along the dotted line, known as the* ecliptic, *by a small amount every day in a direction* against *the daily rotation, as shown by the small arrow next to the sun. The motion of the sun against the celestial sphere is also shown in Figure 3.*

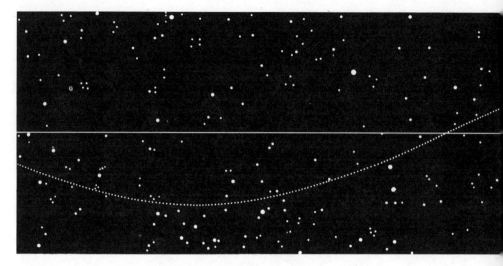

Figure 3. *The motion of the sun against the background of the fixed stars. A narrow strip of the celestial sphere is shown here in the form of a rectangle. As with any such strip on the celestial sphere, the left-hand edge joins onto the right-hand edge. The horizontal continuous line is the celestial equator. The two celestial poles lie outside the diagram at the top (north) and at the bottom (south). The stars in this strip follow a daily motion, with the rest of the celestial sphere, from left to right (east to west). During the course of a year, the sun slowly moves*

After that time, the stars that can be seen in the dawn sky are the same as those that were visible just a year before.

The reason for this time shift is that during the course of a year, the sun moves eastward across the whole sky (Figures 2 and 3). All the stars in the neighborhood of the sun are lost in the bright area of daylight sky surrounding it. A star that was initially in the evening sky gradually remains longer and longer in the daytime sky, where it cannot be seen. But, as the sun travels across the sky, it leaves the star behind, and so the star reappears in the morning sky. During its yearly course across the heavens, the sun is highest in the sky (for those of us who live in the northern hemisphere) in the constellations of Taurus and Gemini. This is our summer. But when its yearly movement takes it into Scorpio and Sagittarius, it is very low in the sky at midday, during our winter. Both the daily height of the sun above the horizon at midday and the appearance and disappearance of stars in the morning and evening twilight reveal the position of the sun in its path across the sky, so the visibility of stars at dawn gives a good indication of where the sun is, and thus of the season.

Ancient peoples were able to create a primitive calendar on this basis. Although the rhythm of the year was shown by the seasons, which deter-

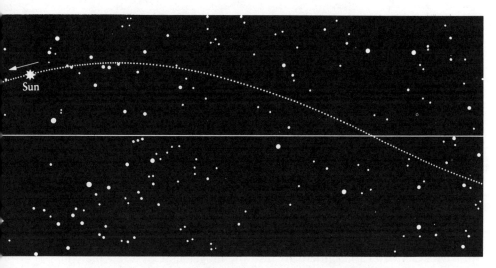

in the opposite direction along the dotted line, i.e., the ecliptic. The sun is at its northernmost point in June, and at its southernmost in December. The arrow shows the direction in which it moves during the year. Each day, it moves by a small amount, while the sphere of the fixed stars, and with it the sun, turns once around the observer. This is the daily motion shown in Figure 1. The half of the celestial sphere in which the sun lies forms the daytime sky above us. The stars in this half of the sky are lost in sunlight and are therefore invisible.

mined people's lives more than they do now, the seasons do not always arrive regularly. It is not always April just because the snow has vanished from the foothills. In ancient Egypt, agriculture needed a more precise indication of the time for sowing, because it was governed by the Nile floods, which occurred regularly every year. A few weeks before this annual event, Sirius could be seen in the morning sky for the first time that year. This was always around 19 July, in our current reckoning. The ancient Egyptians therefore hailed Sirius as heralding the Nile floods and began their year on the day on which this star appeared.

Observation of the first appearance of stars at dawn, even with just the naked eye, allowed events in the sky to be measured quite accurately. Until recently, astronomers used a cross hair in the field of a telescope to determine exactly when a star was due south, obtaining the time of the event from a chronometer. The cross hair used by Babylonian and Egyptian priests on their observation platforms was the eastern horizon, where they noted the first appearance of stars during the year. This procedure was not very accurate, because it is not possible to see a star immediately when it appears above the horizon. It only becomes visible through the haze somewhat later. Therefore, the first appearance of Sirius sometimes occurred a few days before 19 July, sometimes a few days after. Despite the limitations, the Babylonians, in particular, had a surprisingly detailed knowledge of the movements in the sky and made similar records about the same time as the Egyptians. They knew that the sun did not take exactly 365 days to travel across the sky, but rather 365¼ days. Therefore, they added an additional day (a leap day) every four years, just as we still do.

▪ The Planets

The sun moves across the whole sky in the course of one year, but the moon does the same in only one month.[1] In addition to the motions of the sun and the moon against the sky, and the overall movement of objects in the sky that we have described earlier, there are also the movements of a few individual "stars." Some of these are among the brightest in the night sky, and like the sun, the moon, and all the other stars, they rise and set in the usual daily rhythm. However, like the sun and the moon, they move relative to the sphere of fixed stars. As with the sun, this slow motion only becomes evident after a few weeks. The Greeks called these moving bodies, as well as the sun and the moon, *wandering stars*. They are bodies that have no fixed place in the sky. But whereas the movements of the sun and the moon against the background sphere of stars are essentially simple, those of the other wandering stars show hardly any regularity. Even before any written records existed, the remarkable movements of these stars (which were later called planets) must have been noticed (Figure 4). Even among the Babylonians, they were an important subject of astronomical investigation and imbued with particular significance because of their puzzling paths.

The bright star that occasionally illuminated the night sky with its white light is called Jupiter today. There were also the somewhat fainter, yellowish Saturn, red Mars, and the two stars that, despite their apparently irregular movements, stayed close to the sun: Mercury and Venus. Indeed, Mercury is so close to the sun in the sky that it is only visible in the morning or evening twilight. The ancients had known about Mercury for a long time, but Copernicus is said never to have seen it himself. From what we know of Copernicus, however, he was not a particularly dedicated observer. The brighter Venus also remains close to the sun. Although it is not quite as dependent on the sun as Mercury, it still remains in its vicinity, sometimes rising before the sun, and is therefore known as the morning star. At other times, it lags behind the sun and sets after it, appearing as the evening star at dusk, before it vanishes below the western horizon.

The sun and the planets move within a band on either side of the celestial equator (Figures 3 and 4). The sun never appears close to the celestial poles,

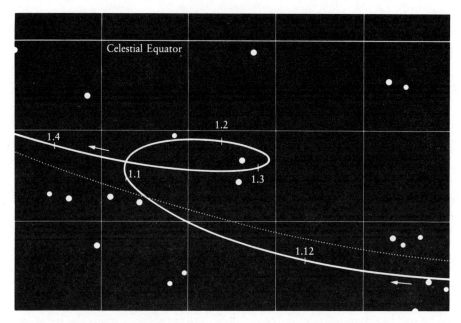

Figure 4. *The planet Venus follows a complicated, looping path over a period of months. The ecliptic (see Figures 2 and 3) is again shown by the dotted line. The celestial equator is the horizontal line in the upper part of the diagram.*

and no planet ever appears so near one of the poles that it fails to rise or set. Observation of the rising of the planets in the morning twilight showed the Babylonians the irregularities in their movements. But not everything was completely irregular.

▪ *Eclipses*

The Babylonians also recorded eclipses of the sun and the moon. They knew that both the sun *and* the moon were involved in both types of eclipse, because solar eclipses occur only at new moon, and lunar eclipses only at full moon. It soon became evident that the pattern of eclipses repeated itself after about 18 years, which still applies today. Using eclipses visible from Germany as an illustration, at the beginning of this decade, for example, there was a partial eclipse of the moon on 17 July 1981, and two total lunar eclipses on 9 January 1982 and 6 July 1982 (only the first of these was visible in Germany). Eighteen years and eleven days later, the same set of events will occur. On 28 July 1999, there will be a partial lunar eclipse, and total lunar eclipses on 21 January and 16 July 2000. The next total solar eclipse visible over the North

American continent will be on 21 August 2017. In the cycle, it is preceded by the total solar eclipse 18 years and 10 days before, on 11 August 1999, which will be visible in Europe. This 18-year cycle of lunar and solar eclipses, known as the *Saros cycle,* is generally reliable. The Babylonians discovered it, and the Greeks used it, but neither knew exactly how the sun and moon moved individually. Thus, Columbus's ability to predict a solar eclipse by using the tables by Regiomontanus does not appear particularly surprising, even though neither of them knew whether the earth and the moon moved around the sun, or the sun and moon around the earth.

▪ The Forgotten Solution

After the Babylonians, the Greeks tried to unify everything into a consistent world picture. There could be no doubt about the stars being part of an unchanging sphere: Anyone could see that with the naked eye on any night. But what part did the wandering stars play, that is, the five planets and the sun and moon, some of which (the sun and moon) moved across the sky more or less regularly, and some of which showed completely irregular motions?

Some Greek scholars thought that the earth was a giant flat disk, surrounded on all sides by the ocean. Others already knew that it must be a sphere; and indeed, they seem to have known more than was once generally thought. Herodotus, who lived in the fifth century B.C., wrote about tribes that lived far to the north and slept six months of the year. Did he have some idea of how the sun illuminated the earth, thus arriving at the idea of the polar night?

The school founded by the mathematician Pythagoras, whose theorem every student still has to learn, had a splendid, modern picture of the world. The earth was a sphere hanging in space and rotating around its own axis. As it turns within the sphere of the fixed stars, we see the stars rise and set. The fixed celestial poles are the directions in which the earth's axis points. This is why they do not move. The sphere of the fixed stars does not turn around us, but we rotate within it. The wandering stars, however, move between the sphere of the fixed stars and the surface of the earth.

The intellectual atmosphere of the Pythagoreans produced a mind of genius: Aristarchos. He has been described as the Greek Copernicus, because he arranged the heavenly bodies in a different, original manner within the sphere of the fixed stars. In the center was the sun, the great central fire, and around it moved the other wandering stars. Mercury and Venus orbited around the sun; Earth came in third place. The Earth, which previously had been thought to be the center of the universe, suddenly became a mere slave of the sun. The Earth rotated on its axis and therefore caused the stars on the sphere of fixed stars and the other wandering stars to rise and set. The moon followed a path around the Earth. Farther out, closer to the sphere of the fixed stars, Mars, Jupiter, and Saturn were all following large orbits around the sun. The system introduced by Copernicus 1800 years later had already been described by Aristarchos.

Like Pythagoras, he was probably born on Samos about 310 B.C. Only one of his writings, on a completely different subject, has come down to us. Aristarchos' great achievement concerning the solar system is known to us only by hearsay. Archimedes, in describing a related subject, writes that Aristarchos of Samos maintained "that the fixed stars and the sun remain stationary, and that the earth travels in a circle around the sun. . . ." This inspired discovery completely failed to influence subsequent human thought on the nature of the universe.

▪ Plagued by Circles

Other intellectual giants overshadowed Aristarchos: namely, Plato and Aristotle. Their ideas had such force that they determined human thought for the next 1000 years, although for astronomy, it was a retrograde step. Plato was no great lover of astronomy and from time to time made caustic comments about it. He was convinced that heavenly bodies could move only on circular paths. The circle was the most perfect figure, and the sphere the most perfect solid.

Aristotle, who influenced Western thought more deeply than any other, summarized the picture of the universe held in his time. It is depicted schematically in Figure 5. In the center of the universe is the Earth surrounded by seven crystalline spheres that enclose one another like the layers of an onion. The innermost sphere carries the Moon. Then Mercury, Venus, and the Sun follow in sequence, each attached to its own transparent sphere; Mars, Jupiter, and Saturn occupy the last three spheres inside the sphere of the fixed stars. The ninth sphere, outside the sphere of the fixed stars, was imagined by Aristotle to be the domain of God, who ultimately maintained the motion of the spheres.

Although this world system had great appeal to many people, it was really only a vague picture. It was no more than a source of discussion over whether the region below that of the moon—the one containing the earth and its atmosphere—was now the highest, most noble location in the universe, or whether such pre-eminence should be accorded to the region outside the sphere of the fixed stars, to which God had been relegated, thus no longer influencing the world from within, but from without.

A world picture such as Aristotle's is not of much value for astronomy if it is unable to explain how the complicated motions of the wandering stars are

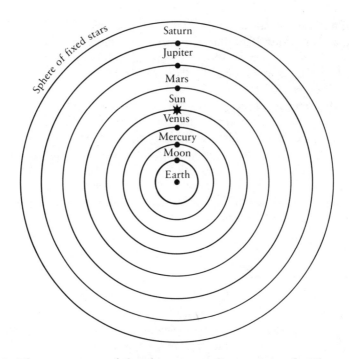

Figure 5. *The movements of the planets according to Aristotle. The sun, moon, and planets revolve around the earth. As Mercury and Venus always remain close to the sun in the sky, they must perforce orbit the earth with the same average period as the sun. The orbits of the wandering stars are enclosed within the sphere of the fixed stars.*

caused. In that respect, it performed badly. Even the sun, which follows the most regular path against the sphere of the fixed stars, moves across the sky in a nonuniform manner. Its behavior cannot be explained by means of a regular rotation of the crystalline sphere to which it is attached; to say nothing of the complicated motions of the planets, as represented by Venus's path across the sky in Figure 4. Thus, the scheme of things had to be revised.

▪ A Complex Mechanism

The necessary refurbishment of the Aristotelean world system took place in the second century A.D. The Greek astronomer Ptolemy described the result in his work *Almagest*. The obviously irregular motions of the wandering stars on the sky had to be explained; yet, Plato and Aristotle must be correct, and only motions in circular paths were permissible. Although it is not easy to reproduce the backward and forward loops in the paths of the planets from circular motions, it can be done. If Jupiter is located on a crystalline sphere with the earth as its center, and if its sphere rotates uniformly, we must see it move uniformly against the background of the fixed stars.

Let us now imagine a second, smaller sphere, whose center is fixed to Jupiter's sphere, and which itself rotates, locating Jupiter on this smaller sphere. The planet can then simultaneously rotate around the earth with the large sphere and be subject to the effect of the rotation of the smaller one. Its motion certainly does not remain regular. At times, the two rotational movements will be in the same direction, and as seen from the earth, the planet will appear to move rapidly across the sky. At other times, the direction of rotation of the small sphere will be the reverse of that of the larger, and the two motions will oppose one another. The planet will move more slowly against the background of the stars, cease to move at all, or even move backwards. An example is shown in Figure 6. It had been known for a long time that irregular motion would result from the superimposition of two regular circular motions that would appeal to any Platonist. The looping motion could be explained on this basis. But, unfortunately, the finer detail hid a catch.

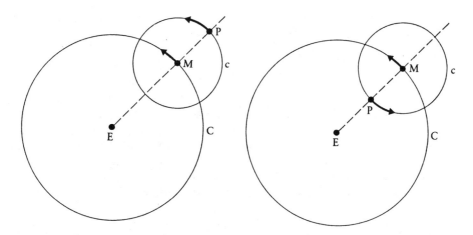

Figure 6. *The movements of the planets according to Ptolemy. On a large crystalline sphere C centered on the earth at E, lies the center M of a small crystalline sphere c, on which the planet P is located. In the left-hand diagram, the motions of the spheres follow the same direction. As seen from the earth, the planet moves rapidly toward the left. In the right-hand diagram, the motions are opposed. The planet moves toward the right.*

As soon as anyone tried to explain the movements of all the heavenly bodies in a quantitative manner, it was generally no longer sufficient to have only two transparent spheres for each wandering star. A third sphere had to be introduced on which the heavenly body was located and which was now being carried along by a threefold rotation of the spheres, in triple circular paths. If its path still did not agree with observation, then these "heavenly ball bearings" had to be made even more complex.

Ptolemy's machinery had yet another fault that marred its beauty. If we imagine a set of spheres whose centers lie on the surfaces of others that are themselves centered on the Earth, the spheres must intersect one another. If one really believes in crystalline spheres, then the wandering stars must, perforce, occasionally pass through the walls of a sphere. Thus, in the Ptolemaic system, the crystalline spheres and subspheres were not true bodies, but simply imaginary geometrical surfaces on which the planets were carried. There was no explanation of why the planets moved in this manner; some sort of driving mechanism was assumed to act upon the planets, a mechanism that could not be understood. This was typical of the pattern of thought at that time. The stars did not belong to our world: They did not obey the laws governing the bodies that formed part of the earthly world. They had nothing in common with anything that existed on the world beneath the orbit of the moon.

▪ The Nature of the Stars

People had not always thought this way. Fifty years before Ptolemy, the Greek Anaximander had pondered the nature of the stars. To him, the heavenly bodies originated in circular, closed tubes in the air that were filled with fire. At various points on the tubes, fire could be seen through tubular openings. Periodically, these openings were blocked, and then became free again. This gave rise to eclipses and to the phases of the moon. Nowadays, we are tempted to laugh at this theory of heavenly blockages. We have no real grounds for doing so, however. It was an immense step forward in human understanding! Anaximander explained the stars as objects that had their origins in earthly things: in fire, tubes with holes, and obviously some other material that could block the holes. That his solution later proved to be incorrect is unimportant. What matters is that he was the first to pose the correct question. Can the world of the stars be likened to the world surrounding us on the earth? Might the heavens not consist of material similar to that found on earth? Do the earthly laws of nature perhaps also prevail in the heavens? Anaximander's ideas can be seen as the forerunners of modern astrophysics, which tries to explain even the most distant heavenly bodies as governed by the laws of physics that have been determined in our own earthly laboratories.

We should also be wary of laughing at Plato and Aristotle, who simply maintained that the wandering stars must move in circles, because circular motion was the most perfect form of motion. Let us be careful: Our modern physics is occasionally not that far removed from this idea.

When physicists endeavor to test mathematical models in elementary-particle physics — the field in which we now seek the most fundamental secrets of nature — it seems that those models that best satisfy certain principles of beauty are the most satisfactory. They should have certain properties of symmetry — the more the better — and they should posses certain so-called invariants. This has been shown to apply in modern physics, but it does not imply that the truth is always expressed in such a manner. The quest for symmetries and invariants is not so different from the Platonic desire to explain the universe in terms of the most symmetrical of all motions, that of a circle.

▪ *Correct Predictions with an Incorrect Mechanism*

The Ptolemaic world system, like the teachings of Aristotle, prevailed throughout the middle ages. People believed in it, and even at the time of Regiomontanus, people had, to some extent, correctly predicted the positions of the planets by using it. This is surprising, because one would think that with an incorrect order of the planets, it would be quite impossible to make any correct predictions about their positions. But this is not the case, because the true motion of the planets is actually extremely simple.

We now know that the planets move in elliptical orbits around the sun, and elliptical orbits can be fairly simply approximated by the superimposition of circular orbits. Columbus succeeded in his trick on the Indians ultimately because the earth and the moon actually follow very simple paths in space, which can be represented quite closely by circular motions. Unfortunately, this is not true for all motions in the universe, nor for all the changes that occur. The weather, for example, is not so obliging. People have tried to express the changes in the weather as regularly recurring cycles, in the hope of being able to make reliable forecasts, without fully understanding the true, complicated, processes taking place in the earth's atmosphere. But no cycle other than that of the seasons has yet been found. Weather forecasting will therefore improve only when we come to a better understanding of the physical processes involved.

With regard to the motions of the planets, it was different. Not only could reliable predictions be made from an incorrect picture of the motion of the wandering stars, but even the new tables, which Copernicus calculated after he had recognized how the wandering stars truly moved, did not reproduce the planets' true paths on the sky any better than those based on the Ptolemaic system.

▪ The Great Change

Much about Copernicus remains puzzling to the present day. Why did Copernicus hesitate so long before he published the truth? Was it fear? Why did he cross out the name of Aristarchos in the manuscript of his masterpiece *De Revolutionibus (Concerning the Revolutions of the Heavenly Spheres)* when he explained that he had come to the solution through studying the Greek philosophers and when it was Aristarchos who, 1800 years before, had arranged the wandering stars in space in exactly the same way as Copernicus did later?

Whatever the circumstances, Copernicus's book appeared in 1543 and barely reached him on his deathbed.[2] In this work, Copernicus explains the motions of the wandering stars by means of a system of circular paths that are no longer centered on the earth but on the sun (Figure 7). The center of the Earth is no longer the center of the universe around which everything turns; it is only the center of the earth's gravitational attraction and the center of the moon's orbit. Mercury and Venus follow paths inside the earth's orbit around the sun: They are called the *inner planets*. Mars, Jupiter, and Saturn follow circular paths outside that of the earth and are therefore called the *outer planets*. The rotation of the celestial sphere with its fixed stars is only apparent and results from the rotation of the earth about its own axis. The universe was again as it had been at the time of Aristarchos. However, Aristarchos probably only showed how the individual wandering stars moved, whereas Copernicus set himself the task of explaining in quantitative terms that in his system the looping paths of the planets could be accounted for quite naturally.

Let us first take a look (Figure 8) at the apparent motion of the centrally located sun. Because of the orbital motion of the earth, we see the Sun as moving in a circular path, the *ecliptic*, across the sphere of the fixed stars. For this reason, the earth's orbital plane is also called the ecliptic. As shown in Figures 9 and 10, it was now apparent how the simultaneous motion of both the earth and the other planets produces looping paths. This world picture sounds so overwhelmingly simple, yet it also posed problems. Copernicus

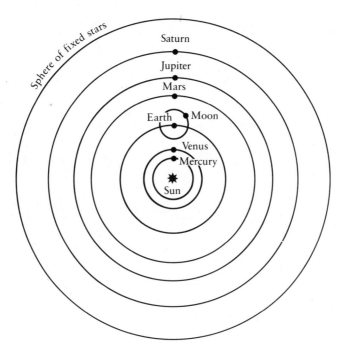

Figure 7. *The motions of heavenly bodies according to Copernicus. The planets, including the earth, orbit the sun, and only the moon revolves around the earth. As in the Greek view, the universe does not exist beyond the sphere of the fixed stars.*

Figure 8. *In the Copernican world system, it appears from the earth as if the sun moves once across the sphere of the fixed stars in the course of a year. This explains the motion of the sun relative to the fixed stars, as described in Figure 3, in a natural manner.*

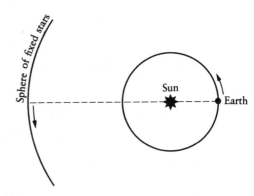

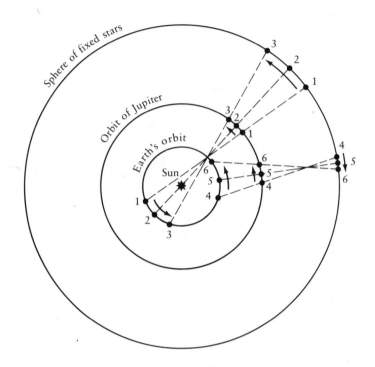

Figure 9. *The motion of an outer planet across the sky according to Copernicus. When the earth is at positions 1, 2, and 3, then Jupiter is at positions 1, 2, and 3 on its orbit. From the earth, it therefore appears consecutively at positions 1, 2, and 3 on the celestial sphere and seems to move in the direction shown by the arrow. But when the earth is at positions 4, 5, and 6, and Jupiter is at positions 4, 5, and 6 in its orbit, it seems to be at the points 4, 5, and 6 in the sky; it therefore appears to move in the opposite direction. To an observer on the earth, this gives the impression of the backward and forward loops of Jupiter's path.*

was still bound by the ideas of the ancient Greeks and did not consider anything other than circular orbits. In his theory, the planets were on rotating spheres that circle *around the sun;* or, more precisely, on rotating spheres whose centers were themselves located on spheres that rotated around the sun. Spheres had to be set inside other spheres. The Copernican world picture was not really simpler than the Ptolemaic one. On the contrary, compared with the later, improved, geocentric system requiring 40 spheres, Copernicus needed 48. As before, the question of the nature of the stars was not considered. The spheres moved both within and through one another, so they were obviously unrelated to the earthly world and its natural laws. As the title of his principal work indicates, Copernicus was only concerned with the motions of the heavenly bodies. For him, it was sufficient to relate the

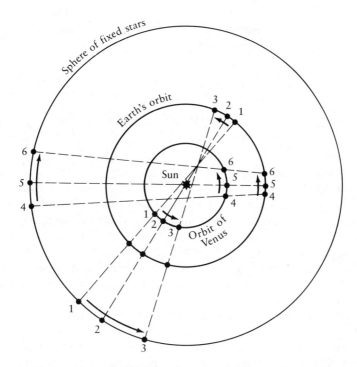

Figure 10. *The motion of an inner planet according to Copernicus. Just as for Jupiter in Figure 9, Venus is observed to move across the sky in one direction at positions 1, 2, and 3, and in the other direction at positions 4, 5, and 6. This, and the fact that the orbits of earth and Venus do not lie in exactly the same plane, causes the loops in Venus's path across the sky, as shown in Figure 4.*

movements of the wandering stars to simpler, circular motions; and the stars were still carried by the outermost sphere, as they were in the Greek model. But now the earth rotated on its axis, causing the stars to rise at the eastern horizon and set at the western, without changing their relative position. Because the earth moves along its orbit around the sun within the sphere of fixed stars, the stars on the sphere should be seen in varying perspective in the course of a year, and so the constellations should vary in shape during this time. Since not even the most precise measurements detected such variations, it had to be assumed that the sphere of the fixed stars was so large that it did not make any difference at all from what point on the earth's orbit it was observed. In the Copernican world system, the radius of the sphere of the fixed stars had to be much larger than that of the earth's orbit. The fixed stars were thus relegated to very great distances. It was only a small step to drop the idea that they were fixed to a physical sphere and to think of the stars as not being rigidly fixed with respect to one another.

▪ Refinements to the Copernican System

Two men were responsible for the next steps: Johannes Kepler, a Swabian, and Galileo Galilei, an Italian. Johannes Kepler (1571–1630) analyzed the observations made over many years by the Danish astronomer Tycho Brahe (1546–1601). The accuracy of Brahe's measurements of the positions of the planets relative to the fixed stars was much better than that of his contemporaries. His *mural quadrant,* built in a north–south direction, was famous. Through the sight of this greater-than-man-sized protractor, the planets could be accurately observed when they crossed the north–south line. The mural quadrant measured the height above the horizon, and a clock gave the exact time of the transit.

In analyzing Brahe's measurements of Mars, Kepler discovered that the planets did not move in circles or in any form of closed paths, other than in ellipses. The sun was not at the center of the ellipse, but at one of the two foci (Figure 11). As a result, the distance between the sun and the planet

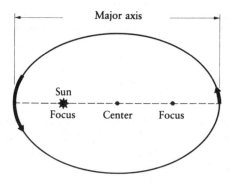

Figure 11. *An ellipse has two foci, which do not coincide with its center. The diameter through both foci is the major axis of the ellipse. Kepler's first law states that the planets move along elliptical paths. The sun is not at the center but at one of the two foci of the ellipse. Kepler's second law describes the velocity of a planet in its orbit. The closer it is to the Sun, the faster it moves along its orbit. The different velocities at the closest and farthest points from the sun are indicated by the arrows of different lengths. The elliptical orbits of the planets are far less elongated than shown in this diagram. They closely resemble circles (see, for example, the planetary orbits shown in Figure 78).*

varied throughout the orbit. The fact that the planets move in elliptical orbits is known as *Kepler's first law*. *Kepler's second law* describes how fast the planets move at different points in their elliptical orbits. In fact, Kepler found that they move faster when they are closer to the sun than they do in the more distant parts of their orbits. Finally he showed that for any two planets, the one with the larger-diameter orbit required longer to complete one revolution around the sun. The simple law governing this is *Kepler's third law*, shown in Figure 12. It not only states that the planet with the larger-diameter orbit has a longer orbital period, but indicates how the orbital period may be calculated exactly from a knowledge of the orbit's diameter. Because the planets move in elliptical orbits, we must explain what is meant by an orbit's diameter. The law relates to the elliptical orbit's *major axis*, which is defined in Figure 11. The major planets are shown in Figure 12, together with Halley's comet, which also has an elongated elliptical orbit (see Chapter 11, Figure 78), but which still obeys Kepler's third law.

Galileo Galilei (1564–1642), almost universally known simply as Galileo, came from Pisa and was only slightly older than Kepler. Although Kepler

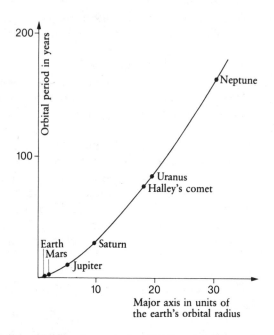

Figure 12. *Kepler's third law is a relationship between the major axis of the orbit (Figure 11) and the orbital period. All bodies in elliptical orbits around the sun fall precisely on the curve indicated.*

repeatedly tried to correspond with him, Galileo hardly ever answered his letters.[3]

Galileo was convinced of the correctness of the Copernican view, but initially refrained from publishing his comments for fear of conflict with the church. When he directed his telescope—which had been invented in Holland, and a version of which he had built himself—toward the heavens, he not only saw that the face of the moon had mountains, but discovered four moons orbiting Jupiter. Here he saw a system, consisting of Jupiter and its moons, similar to the Copernican picture of the sun and the planets: a sort of miniature solar system.

It is often said that scholars of his time refused to look through Galileo's telescope. They stubbornly clung to their belief that the moons of Jupiter could not exist. A look through the telescope would have convinced them otherwise.[4]

We should realize that Galileo's telescope did not have the quality of even the cheapest, store-bought modern telescope. With this in mind, it is not surprising that he had difficulty in convincing his opponents with the evidence of their own eyes. One wit even said that Galileo's greatest achievement was not in discovering the satellites of Jupiter, but in actually being able to find Jupiter in the field of his rickety telescope. It was certainly necessary to take a very careful look through the telescope in order to be certain that the faint points of light close to the bright disk of Jupiter were not reflections in the telescope. Naturally, Galileo had checked that, but it was not easy to demand equal care from his opponents.

Through his telescope, Galileo did not see Venus as a circular disk. It showed phases like those of the moon. At one time it was a semicircular disk, at another full, and at yet another just a narrow crescent. However, if Venus moves around the sun and is illuminated by it, this should be obvious (Figure 13).

In the Ptolemaic system, the phases of Venus could not be explained. Let us look again at the Greek world view shown in Figure 5. There Venus had to lie somewhere between the earth and the sun, year in, year out; and in order to remain in the vicinity of the sun, it was forced to orbit the earth with the same period as the sun. Two possibilities existed: Either Venus itself shone, in which case it would always be "full," or we only saw it by reflected sunlight. But then it would always appear as a small crescent in the Ptolemaic system, because, as seen from the earth, the sun was always behind it. Modern photographs clearly show the phases of Venus that Galileo discovered (Figure 14). This discovery was a convincing argument in favor of Copernicus's teachings.

Most of Galileo's discoveries with the telescope were not disputed by his clerical opponents. On the contrary, people were convinced and urged him to make further observations. Neither the moons of Jupiter nor the phases of Venus were disputed; only the Copernican system was rejected.

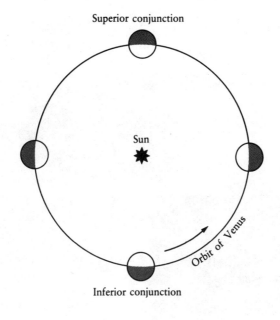

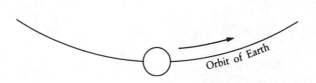

Figure 13. *The phases of an inner planet in the Copernican system. The hemisphere of the planet illuminated by the sun appears sometimes as "full" and sometimes as a crescent (see also Figure 14). If the planet is between the sun and the planet, it is said to be at* inferior conjunction. At superior conjunction, *it is on the other side of the sun to the earth. At inferior conjunction, we are looking at its dark side, and at superior conjunction, it appears fully illuminated.*

Galileo was a convinced Copernican, but he still believed in circular orbits, being sure that the circular orbit of any body corresponded to its natural motion. He believed that any object that was free from external influences would move in a circle just like the planets. Kepler, on the other hand, thought that the planets required some driving force to keep them moving along their elliptical orbits. He visualized some force that originated in the

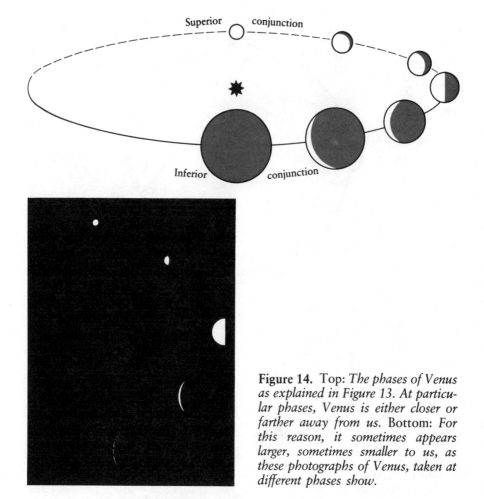

Figure 14. Top: *The phases of Venus as explained in Figure 13. At particular phases, Venus is either closer or farther away from us. Bottom: For this reason, it sometimes appears larger, sometimes smaller to us, as these photographs of Venus, taken at different phases show.*

sun, and which swept the planets around the central star like some gigantic broom. Both were wrong; nevertheless, both were close to the truth.

As we run through the various ideas mooted in the first half of the seventeenth century, we, who know the answers, are able to see the Saint Elmo's fire that foretells the flash of lightening that will clear the air. The explosion was started by just one man, who was born in England a year after Galileo's death.

TWO

Planetary Ballistics

One thing I cannot omit here, and indeed I thought it was extraordinary, as least it seemed a remarkable hand of Divine justice, viz., that all the predictors, astrologers, fortune-tellers, and what they called cunning-men, conjurers, and the like; calculators of nativities and dreamers of dreams, and such people, were gone and vanished; not one of them was to be found. I am verily persuaded that a great number of them fell in the heat of the calamity, having ventured to stay upon the prospect of getting great estates; and indeed their gain was but too great for a time, through the madness and folly of the people.

DANIEL DEFOE (1660–1731)
Journal of the Plague Year

What is the most natural motion for a body? At first sight, it might seem that a body's natural tendency is to revert to a state of rest as soon as possible. A golf ball finally comes to rest somewhere or other, however strong the original stroke that set it in motion. A sailing ship merely bobs up and down on the water without going anywhere when there is no wind. It would appear that any body is forced to become stationary at the earliest possible opportunity. Only the planets tirelessly follow their orbits round the sun, year in, year out. Are celestial bodies — such as the moon, for example — quite unlike the earth itself? But anyone who looked through the newly invented telescope saw that there were mountains on the moon, resembling features found on earth.

▪ *The Birth of Force*

We can see this dilemma in Galileo's work. He set down his explanation of motion in four discussions in form of dialogues involving three men. Galileo depicted himself as a person called Salviati. Then there was Sagredo, an intelligent layman, who professed to be nonpartisan, but who finally always supported Salviati-Galileo. The third was called Simplicio. Although Galileo insisted that he had no ulterior motive for using this name, Simplicio always turned out to be the "fall guy."

The first of these dialogues, which altogether stretch over four days, discusses the similarity between the earthly and the cosmic worlds. We recognize here the fundamental scientific issue of his day. Galileo had realized that the motion of earthly bodies is strongly influenced by friction. A ball thrown through the air is braked by air resistance and, when it reaches the ground, by friction with the surface. This brings it to a stop. If friction is reduced, then the body takes longer to come to rest. A heavy stone, resting on an icy surface, travels a long way when it is given a push, because the friction is very low. On a patch of grass, where the friction is high, the stone stops as soon as one ceases pushing. The apparent inclination for a body to revert to a state of rest is only a result of ever-present friction. It was already clear to Galileo that a state of rest was not the natural state for material bodies free from friction. But what would the motion of a body be if it were left to its own devices and not subjected to any external influences, such as friction? Although Galileo was firmly convinced by Copernicus's world picture, he was still strongly influenced by Aristotle and Plato. Had not Copernicus himself believed in the Greeks' sphere of the fixed stars that was the outer limit of the universe? Does that not suggest that the natural motion of bodies is to move in a circle? Would not motion in a straight line inevitably lead to bodies colliding with the sphere of fixed stars? If the planets are left to their own devices, they must move in circular orbits, because it is the most natural type of motion. The circle was the most perfect figure and had therefore been chosen by the Creator for the paths of the planets—Galileo had read his Plato. He was a dilatory correspondent with Kepler, and had never paid any heed to Kepler's elliptical orbits.

It was not everyone's view that the circle was the most natural path in which bodies would move. In France, René Descartes (1596–1650) maintained that, left to itself, a body would continue in a straight line, just as an arrow in flight follows a straight line (or at least a nearly straight line, since it is not completely free from other influences; its weight pulls it downward). Descartes should have received the credit for this suggestion, although it is usually credited to Newton. Newton, however, was still learning to read and write when Descartes died. Descartes was the first to understand that of all possible paths, motion in a straight line was the "natural" one, despite the fact that bodies on the earth did not move in straight lines. We now know that a body out in space, far from any heavenly bodies that might influence it, retains its initial direction of motion. If it starts out in a particular direction, it will continue to move in that direction forever. This was all the more difficult to accept, because no one ever saw bodies that were moving without any external influences. Bodies on the earth were soon reduced to their "natural" state of rest as a result of friction. The moon is influenced by the earth and is not moving freely; the planets are constrained by the sun to move in their orbits.

Galileo and Descartes were the first to see that bodies can move on their own and do not have to be continually acted upon to remain in motion. They were ahead of Kepler, who believed that a force emanated from the sun and kept the planets in motion. This, like a broom, swept them round in their orbits and prevented them from coming to a stop. To Galileo, motion in a circle was natural; it required no influence from outside and was self-sustaining. To Descartes, motion in a straight line was natural, and only additional influences from the earth and the sun caused bodies to deviate from their linear paths.

So far, we have always carefully referred to "influences" of other bodies. What can cause a body to deviate from its "natural" path? The nature of *force* was already understood. It was felt by the muscles when one tried to get a bogged-down wagon moving. People knew that a force could be transmitted from one body to another with levers and ropes. A horse pulls against the harness and thus exerts a force on it, and the harness on the cart. Thus, the horse is pulling the cart without being physically in contact with it. In addition, people knew about the mysterious force of magnetism, which operates through empty space. The tendency for all bodies to fall downward whenever they are free to do so seemed to be the result of a similar force. Instead of speaking of the motion of a body unaffected by any other bodies, it was better to speak of its motion when no forces were at work on it.

▪ The Cannon on Top of a Mountain

We do not know exactly how Isaac Newton (1643–1727) arrived at his revolutionary ideas. In general, he just gave his results, but left the reader in the dark as to how he arrived at them. He later wrote that most of his discoveries concerning the motion of the heavenly bodies, and above all, the law of gravitation, were made in the years 1665 and 1666. He was then 23 years old. The plague was sweeping England—in London alone almost 30,000 people died in the epidemic during the summer of 1665. Daniel Defoe, a contemporary of Newton, later gave a full description of events in

London during the plague year. People fled from the towns into the country, where the smaller population was thought to lessen the danger of being infected.

With the plague raging, Cambridge University, where Newton was studying, had to curtail courses. Newton went back to his place of birth, Woolsthorpe near Grantham in Lincolnshire. In the tranquillity of country life, ideas pointing the way ahead seem to have come to the young man, ideas that he was, in part, to follow up and publish only decades later.

Let us try to summarize the scientific beliefs of his time. People knew about gravity, the force that dragged a stone down to the ground. Galileo had discovered the laws governing the motion of falling bodies, laws that allowed the path of a shot fired from a cannon to be predicted. Then there were the elliptical paths of the planets, which were somehow held in their orbits by a force emanating from the sun. A similar mysterious force seemed to be holding the moon in its orbit around the earth. It was then, while taking refuge from the plague in his birthplace, the young Newton had the crucial idea: The earth's gravity and the strange force that held the moon in its orbit around the earth were one and the same.

Newton's idea can best be understood by using an example that he himself used later: the idea of a cannon on the top of a mountain. This is shown in Figure 15. Let us assume that on top of a mountain a cannon is pointing horizontally, i.e., more or less at the horizon. If a cannonball is now fired, it will first travel horizontally but is then dragged downward by gravity. It travels along the path marked A, soon reaching the surface of the earth. Let us repeat the shot, but this time giving the cannonball a faster velocity by using a larger charge of powder. Essentially the same thing will occur. The shot will reach the surface along path B. If we continue to increase the velocity of the cannonball, succeeding shots will travel farther and farther, as shown by paths C and D. The fact that the earth is a sphere is really making itself felt. If the gunner increases the velocity of his cannonball even more, he will have to be careful not to shoot himself in the back. The cannonball will no longer fall to the earth's surface but will follow a circular path around the earth. The cannonball has thus become an earth satellite, just like the moon. It must therefore be moving in such an orbit that at any instant the attraction of the earth and the centrifugal force counterbalance one another.

This explanation showed that the orbit of the moon and the motions of falling bodies, which Galileo himself had investigated, were one and the same. The motion of the moon — like the motion of freely falling bodies — is determined by the earth's gravity. The mysterious force that keeps the moon moving in its orbit around the earth is none other than gravity, the same force that we experience throughout our lives. Newton also began to realize what kept the planets moving in their orbits around the sun. Since the heavenly bodies consist of the same material as bodies found on earth, so, like the earth, they must exert a gravitational force. Just as the earth keeps

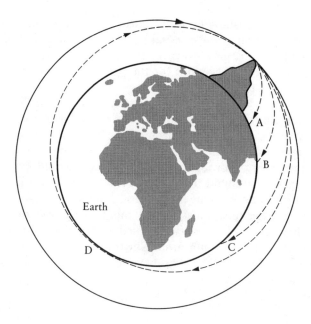

Figure 15. *Depending on its initial velocity, the shot from a cannon on top of a mountain hits the earth's surface at points A to C. If it is given a sufficiently high velocity to reach the antipodean point D, it grazes the surface and returns to its starting point. If it moves even faster, its orbit around the earth may become a circle (solid line).*

the moon in orbit around it, so must the sun retain the planets in their orbits. Did people not already suspect that a force emanating from the sun was acting upon the planets? Had not Kepler assumed that this force declined with distance, because the outer planets moved around the sun more slowly than the inner ones? Kepler's third law, as shown in Figure 12, helped Newton even more.

From the way in which the planets moved around the sun, Newton was able to deduce the strength of the force of gravity close to it, in the region of Mercury and Venus. Mars, Jupiter, and Saturn, however, showed him the strength of the sun's gravity at greater distances. This enabled him to calculate how the sun's force of attraction declined with distance: double the distance, a quarter of the force; three times the distance, a ninth of the force. I have followed Newton's ideas in Appendix A, describing Newton's law of gravity in simple terms.

In discovering the law of gravity, Newton relied heavily on Kepler's third law, which is given in Kepler's book *De Harmonice Mundi* (*Harmony of the Spheres*). In this work, he made a futile attempt to relate the distances of the planets to the relationships between musical notes, an idea that had no relevance to the motion of the celestial bodies. A law stated in the fifth book of this work, however, gave Newton the key to his own law of gravity. Newton was thus able to account for the paths of not just the planets and the moon, but also those of comets.[1]

Newton's law was not brand new. It was already known that the strength of the radiation emitted by a source of light declined in accordance with the same law: twice the distance, a quarter of the brightness, and so on. It had even been suggested that the same law applied to gravity. But Newton was the first to investigate thoroughly how the planets would move if this law was valid in space. He found that Kepler's three laws then followed inevitably, and went on to develop the mathematical methods required to calculate the orbits of the planets. This opened the way for the development of celestial mechanics, which is a self-contained part of modern physics. Only in 1915, a quarter of a millenium after Newton, did Albert Einstein (1879–1955) succeed in taking the first steps beyond the whole edifice of mechanics based on Newton's work.

▪ *The Cannon in Space*

How does a body move under the influence of the gravitational forces of other masses? We can understand this by doing a thought experiment. Let us assume that the universe is completely free of bodies that exert any gravitational attraction. Let us further assume that in this otherwise empty universe, we have a cannon, which we fire in any arbitrary direction. The shot will then continue, for all eternity, to move in a straight line in exactly the original direction, retaining the velocity with which it emerged from the cannon. There is no force, no friction, to alter its path. The only influence would be the attraction exerted by the cannon, together with the gunner. Since both are material objects, they exert a gravitational attraction, which acts on the shot. For the time being, however, we shall ignore this minute effect. Now let us add a star, such as the sun, somewhere in space. Let the cannon be 150 million kilometers from the sun. This distance corresponds to the earth–sun distance. The gun will now be attracted by the sun's gravity and, together with the gunner, would crash into the sun. We can stop this by linking the gun with a rocket, whose thrust tries to push it away from the sun. If its outward thrust is exactly as great as the sun's inwardly directed gravitational attraction, then both forces balance one another, and the cannon remains at the same distance from the sun. How will any shots that are fired behave, now that they are subject to the sun's gravity?

We can envisage a whole confusing host of paths for the shot if we fire in any direction with any conceivable charge. We will, however, restrict ourselves to firing in just one direction, namely, at 90° to the sun, or to the left in Figure 16. We begin with the simplest conceivable case and assume that the shot is fired with minimal velocity, almost as if we had completely forgotten to use any powder, or, to put it another way, as if we had slowly pushed the shot up the barrel of the cannon toward the muzzle with a piece of wood until the cannonball finally appeared from the end of the barrel and lost contact with the cannon. It has, so to speak, been fired with a velocity of zero. At the instant it becomes free in space, it feels the sun's gravitational force, which was previously counteracted by the pressure exerted by the side

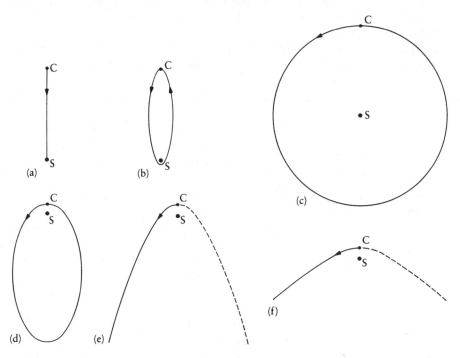

Figure 16. *A shot fired from a cannon C in the sun's gravitational field describes various paths, according to how great its initial velocity was. The individual examples are explained in the text. In parts (e) and (f), the motion follows a parabolic and a hyperbolic orbit, respectively. In these cases, the sun is still at the focus of the curve. With increasing distance from the sun, the hyperbolic orbit approaches a straight line more and more closely. If one imagines the shot as being fired horizontally to the left, then in the cases of the circle and the ellipses, it follows the paths indicated by the solid line. Note that in the lower three cases, the scale has been chosen such that C and S appear closer to one another. The broken lines show that the paths form parts of a full parabola and hyperbola, respectively.*

of the barrel lying closest to the sun. Once free in the sun's gravitational field, the cannonball moves in a straight line toward the sun. The closer it gets, the more it is affected by the sun's gravity and the faster it falls. Its path is a straight line. After $64^{1}/_{2}$ earth-days, the cannonball will reach the sun, where it will be vaporized. But that was not a proper shot. We did not really need a cannon for it. For our next experiment, we will not forget the gunpowder; so, let's load again and fire. However, we will still not let the initial velocity be very large. This time, upon leaving the barrel, the cannonball will again be affected by the sun's gravitational attraction. But now, it will follow a small, elliptical path around the sun (Figure 16b), return to its starting point, and endanger the cannon and the gunner. If we quickly rescue the gun and its crew with our rocket, then the cannonball will follow its elliptical orbit around the sun forever. The sun is not located in the center of this ellipse but at one of its two foci. The orbital period will be rather more than 129 earth-days. If we repeat the experiment with a still greater charge so that the cannonball will leave the gun with a still larger velocity, then the orbit will be a somewhat fatter ellipse, and the orbital period will be greater.

If, in the next trial, we give the cannonball a velocity of 30 kilometers per second, it will take about a year to follow a circular path around the sun (Figure 16c). The sun itself is in the center of the circle.

Ignoring the fact that in our thought experiment we have already far exceeded the velocities that any actual gun can give to a shell, let us increase the charge of gunpowder once again. The orbit will still be an ellipse. But, whereas with the elliptical orbits discussed so far in Figure 16b the gun was at the point most distant from the sun, it is now at the point closest to the sun (Figure 16d). The more the initial velocity increases, the farther out into space the orbit reaches, and the longer it takes for the shot to complete an orbit around the sun. As previously, the orbital periods are determined by Kepler's third law. If one were to plot the major axis of the orbit and the orbital periods on the diagram shown in Figure 12, all the points would be found to fall exactly on the curve. The orbits that have long orbital periods and extend far out into space therefore have large major axes. When the velocity reaches 42 kilometers per second, however, something completely different happens. The cannonball never comes back! The elongated ellipses become a curve, a *parabola,* that stretches to infinity (Figure 16e). The farther the cannonball travels away from the sun, the slower it moves. At that particular velocity, it would finally come to rest at some infinite distance from the sun.

If we make the initial velocity even greater, then the cannonball definitely does not come back. In the case of the parabola, it only barely escaped from the sun's gravitational attraction, because its velocity became zero at infinity. This time, whatever the distance from the sun, it retains a perceptible velocity. Because it is then hardly affected by the sun's gravity, it moves outward essentially along a straight line (Figure 16f). A curve of this sort is known as a *hyperbola.*

However a shot is fired in the gravitational field of a star, its path is a circle, an ellipse, a parabola, or a hyperbola. Only when its initial velocity is zero, or when it is fired directly toward, or away from, the star, does the path become a straight line. If a body falls toward the sun from a great distance, it travels along a very elongated elliptical orbit, appearing as a parabolic or a hyperbolic orbit close to the sun. After a short appearance, it disappears into space again along the same parabola or hyperbola. Anything that arrives from a great distance returns to a great distance. For this reason, a star on its own is not able to capture other bodies and permanently bind them to itself gravitationally; it requires the help of at least a third body. There is a science-fiction story about the inhabitants of a planet that, for some reason or other, had lost its sun. On their planet, they were traveling through frigid space, trying to survive by freezing the whole population for thousands of years, periodically allowing a few astronomers to wake up. The latter had the task of finding out whether, in its odyssey through the Milky Way, the planet would be lucky enough to encounter, in the near future, the gravitational field of a star that would capture the homeless planet, its warmth allowing the people in the freezer to resume an active life. The defrosted astronomers knew their celestial mechanics and ignored single stars, which their planet would merely race past on a parabolic or hyperbolic orbit before returning to the icy depths of space, before even having had the chance to get thoroughly warm. They only kept a lookout for double-star systems, knowing that out of all the paths that a body can follow around a single star, there were none that began at a vast distance and ended permanently close to the star. As far as celestial mechanics was concerned, the story was perfectly correct.

The sun does, however, capture occasional comets that approach it from great distances along almost parabolic paths. It cannot do this alone; if the comet, as it approaches the sun, also happens to come within the gravitational field of a planet, particularly Jupiter, then its orbit may be perturbed into an ellipse, which it thenceforth follows around the sun. We shall return to this point in Chapter 7.

The motion of a body in the gravitational field of a *single* star, such as we have discussed earlier, is simple. The problem of capture leads us to the question of motion in the gravitational fields of two or more bodies.

▪ The Two-Body Case

The thought experiment involving a cannonball in the gravitational field of the sun is a particularly simple form of the motion of two bodies in otherwise empty space, because although the sun attracts the shot, it is itself unaffected by the latter. Those who investigate celestial mechanics call this the *one-body problem,* because only one body moves. In reality, such a simple situation never occurs. The cannonball consists of matter and therefore possesses its own gravitational attraction. In comparison with the sun's gravity, however, one can neglect the gravity produced by the shot. What would happen, however, if in our fictitious experiment, we were able to fire a shot that consisted of as much matter as the sun itself? The gravitational field of the shot would then be the same as that of the sun. How would our cannonball move in that case? Both bodies would be equal partners. The cannonball would no longer move around the stationary sun, but both bodies would orbit one another.

We will now consider a particular, simple case of this otherwise rather complicated motion: orbit in a circle. If the mass of one of the bodies is very small, it can move in a circular orbit around the larger mass. If both bodies have the same mass, then they orbit around a point halfway between them. In general, two mutually attracting bodies orbit a fictitious point lying somewhere between them that is known as the *center of gravity.* If the two bodies are of equal mass, then this point lies halfway between them, otherwise it lies closer to the more massive body. The more the mass of one of the bodies exceeds that of the other, the closer the center of gravity lies to the center of the more massive object. In the case of the sun and a cannonball, the center of gravity essentially lies at the center of the sun. If the cannonball had the mass of Jupiter, however, and orbited at Jupiter's distance, then the combined center of gravity would lie 47,000 kilometers above the surface of the

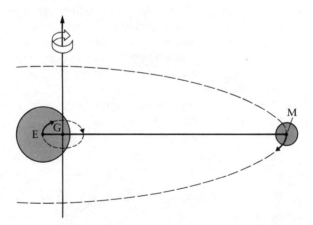

Figure 17. *In solving a "proper" two-body problem, the gravitational force of one of the bodies cannot be neglected, as was done in Figure 16. The two bodies — here the earth E and the moon M (not to scale) — move in elliptical orbits around their combined center of gravity G. Two bodies may also pass one another on parabolic or hyperbolic orbits.*

sun. In fact, the sun and Jupiter do orbit this point. Similarly, the moon does not orbit the earth's exact center, but both bodies orbit their combined center of gravity, which lies about a quarter of one earth radius below the surface of the earth. During the course of a month, not only the moon, but also the earth, orbits this point (Figure 17).

So far, we have discussed only the case of general motion of two bodies in a *circle*. As with the motion of an essentially massless shot (when compared with the sun), there are also elliptical, parabolic, and hyperbolic orbits for bodies of differing mass. Just as in the case of the nearly massless cannonball, where the sun remained stationary at the focus, two comparable masses move such that the combined center of gravity always remains at the focus of the ellipse, parabola, or hyperbola. A circular orbit is thus a special case of an elliptical orbit, in which the two foci are so close together that they form a single point at the center of the circle. The motions become even more complicated when a third body orbits within the gravitational field of two others.

▪ *The Five Points of Joseph Louis Lagrange*

When the gravitational fields of three bodies interact, what paths do the bodies follow in space? This question has exercised mathematicians ever since the law of gravity has been known. It is often said that the problem posed by three bodies has yet to be solved. However, that requires explanation. If we think of any three stars in space, we can, in principle, determine how they move at any particular instant of time, i.e., in which directions they move, and with what velocities. Nowadays, with any small computer, we can even calculate how this triple system will behave in the short term under the mutual influence of their gravitational forces. The so-called insolubility of the "three-body problem" lies in the fact that there is no *exact* mathematical theory for the motion as there is for the motion of two bodies. Mathematicians can deal with two bodies that mutually interact, but not with three. By using computers, however, we can now solve individual cases of the motion of three bodies.

Although a complete solution is still not possible, we do know of some forms of motion that are both simple and also encountered in nature, and more recently, are used by man as well.

In the year 1772, the French mathematician Joseph Louis Lagrange (1736–1813) published a work that won him a prize from the Paris Academy of Sciences. He entitled it "A discussion of the three-body problem." In it, among other solutions, he considered the case in which only two of the bodies had perceptible masses, and the third essentially massless, body moved in the gravitational fields of the other two. A spent rocket stage in the gravitational fields of the earth and the moon might be an example of such a case. The earth and the moon move as if the third body did not exist, but the rocket is completely at the mercy of the gravitational fields of the other two bodies.

In order to understand Lagrange's solution of this simplified example of the three-body problem, let us assume that the two main bodies, the earth and the moon, move in circles around their common center of gravity, as we described earlier. They are like two riders on a merry-go-round, on opposite sides and at different distances from the center. Now imagine that we are somewhere on that merry-go-round. As we are moving with it around its center of gravity, the two bodies appear to us to be stationary; we are rotating with them. As a result, we experience not only the gravitational attraction of the two bodies, but also the centrifugal force caused by the rotation.

There are indeed five points where the gravitational forces and the centrifugal force that we experience on the rotating merry-go-round are in balance. These are indicated on Figure 18 by $L_1, \ldots, L_5$. It is easy to see that there must be one such point on the line joining the two bodies. Close to the earth, gravity predominates, pulling every object down toward the center of the earth. Close to the moon, its gravity acts in the opposite direction. To this we

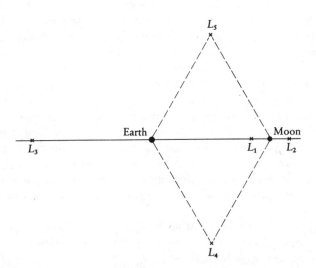

Figure 18. *While the earth and the moon rotate about their common center of gravity, the five points $L_1 \ldots L_5$ experience both gravitational and centrifugal forces. At these points, known as the Lagrangian points, the forces are in equilibrium.*

can add centrifugal force, which in the right-hand half of the diagram is directed toward the right, and in the left-hand half toward the left. Thus, somewhere there must be a point where the forces acting toward the right and those acting toward the left are in balance. This point is marked L_1 in the diagram. On our merry-go-round, a body at point L_1 experiences no forces. Once at that point, it will stay there. On the line joining the earth and the moon (which is naturally rotating with the merry-go-round), there are two other similar points, which lie outside the earth and the moon. Let us take the point marked L_2. Here again, the forces cancel out. The earth and the moon pull toward the left, and the centrifugal force toward the right. Farther to the right of L_2, the centrifugal force is stronger, and to the left weaker, than the combined gravitational forces. Exactly the same applies at point L_3 on the left-hand side of the diagram. But in addition, Lagrange discovered two other points on the rotating merry-go-round where no forces are experienced. These are shown as L_4 and L_5. Let us take L_4 as an example. In the diagram, the gravitational forces of the earth and the moon act toward the top, and centrifugal force toward the bottom. At point L_4, these forces are in balance; the same applies at L_5. The positions of the two points can be found by constructing equilateral triangles, where the earth and the moon are two of the corners, and L_4 or L_5 is the third. Bodies at the five Lagrangian points experience no forces and do not move in the rotating merry-go-round set up by the two main bodies. In the earth–moon system, L_1 lies 326,000 kilometers from the earth on the line joining the earth and the moon. It is only 58,000 kilometers from the moon. Point L_2 lies approximately the same distance away from the center of the moon—behind it, as seen from the earth. Point L_3 lies 379,000 kilometers away on the side opposite to the moon. The distances of points L_4 and L_5 from the earth and the moon are naturally equal to the distance between the two bodies, i.e., 385,000 kilometers.

Is this all just an intellectual game on the part of a French mathematician? Not at all. In the solar system, apart from the sun, the giant planet Jupiter plays a decisive role, despite having only about one thousandth of the sun's mass. But three quarters of the total mass of all the planets is found in Jupiter. It is the chief planet, and its gravity plays a very important part along with that of the sun. We have already mentioned how, together with the sun, it is able to capture comets. The sun and Jupiter form a two-body system that revolves around their combined center of gravity.

Many minor planets—there are countless numbers in our planetary system (see Chapter 8)—can be found at the L_4 and L_5 points created by the sun and Jupiter. These minor planets are known to astronomers as the *Trojans*. Like Jupiter, they revolve around the sun with an approximately 12-year period. They are not found at the precise points that Louis Lagrange described in his prize-winning paper, but very close to them (Figure 19). Thus, the gravitational attraction of the sun and Jupiter and the centrifugal force

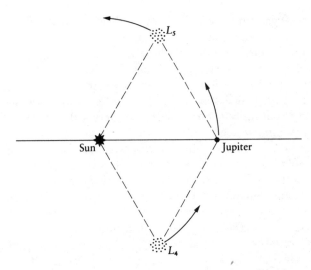

Figure 19. *In the Sun–Jupiter system, there are five equilibrium points, as described for the Earth—Moon system in Figure 18. Close to the L₄ and L₅ points on Jupiter's orbit, minor planets have congregated over time and now revolve around the sun with the same period as Jupiter. These are known as the* Trojans.

are not precisely balanced. A small force still remains that always pulls the Trojan minor planets back to the points L_4 and L_5, however. The bodies oscillate slowly about these points, always remaining close to them.

This applies not only to the Sun–Jupiter system, but has a more general significance. The L_4 and L_5 points are able to retain bodies in their vicinity. If any perturbation tends to cause the bodies to move slightly away from the Lagrangian points, the general gravitational fields of the two main bodies and the centrifugal force act to restore them to their original position. The same does not apply to the L_1, L_2, and L_3 points. If a body at one of those points is acted upon by any force—even if it is only moved a slight distance away from the point—the combination of forces causes it to be pulled even farther away.

The earth and the sun form a system similar to that of the sun and Jupiter. Because of the much smaller mass of the earth—it amounts to only about one-third of 1 percent of the mass of Jupiter alone—it is of far less significance. The combined forces of the earth and the sun are indeed perturbed by the forces of the other planets, but Jupiter is far away. The L_1 point in the sun–earth system lies at a distance of 15 million kilometers away from the earth in the direction of the sun. This point is still quite far away from the orbit of Venus, which crosses the earth-to-sun line at a distance of 41 million kilometers. It also lies far outside the moon's orbit, which is only half a million kilometers from the earth. Even the moon does not perturb bodies at the L_1 point in the sun–earth system. Any body, once it has reached that point, will remain stationary as the sun–earth system rotates, because the gravitational forces of the earth and the sun and the centrifugal force are in balance. But L_1 is one of the Lagrangian points at which bodies are unstable if they are subject to forces that move them even a small distance away. If a body lies at the L_1 point in the sun–earth system, it orbits the sun once a year. No minor planets have yet been discovered at that point, but August 1978, a space probe was launched that was to stay there for years, as a sort of artificial Trojan.

▪ *The Strange Journey of* ISEE-3

ISEE-3 is the abbreviation for *International Sun–Earth Explorer 3*. This U.S.-European space probe was designed to investigate the streams of gas emitted by the sun known as the *solar wind* (see Chapter 7). Equipped with a large number of measuring instruments, it was sent to the L_1 Lagrangian point in the sun–earth system, where it stayed between the sun and the earth in order to measure the solar wind before it encountered the earth. As we saw earlier, the L_1 point is not a very suitable one. Any body that deviates, however slightly, from the exact point, is pulled even farther away from the point by the various forces that exist in the system. As long as it remains close to that point, however, these forces are small, and a small thrust from the rocket motors suffices to keep it at the L_1 point. For four years, *ISEE-3* oscillated about the L_1 point—which Lagrange had postulated more than 200 years before—and measured the activity of the solar wind. After that, it was given a new mission. In the middle of 1982, its motors were fired in order to move it toward the earth. The probe became a satellite of the earth (Figure 20). Its orbit was so far from the Earth, though, that it was strongly influenced by the moon. It also did not have the slightest similarity to the elliptical, parabolic, or hyperbolic orbit it would have had if it had been subject only to the earth's influence. The space probe flew past the earth, and would have nearly escaped into space had the combined gravitational forces of the earth and the moon not forced it to turn back. It again passed close to the earth, and then followed a path like a figure eight lying on its side. It stayed within the earth's magnetic tail (see p. 71) early in 1983. When it crossed the moon's orbit on 30 March, it passed very close to the moon, as had been intended from the start. The earth's natural satellite again sent it out to leeward of the earth. On 30 June 1983, *ISEE-3* reached its most distant point from the earth and then came back in another figure eight. It

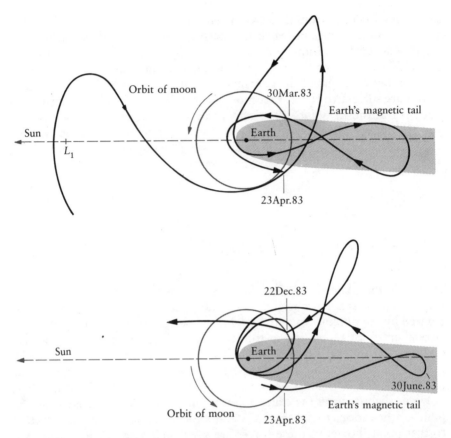

Figure 20. *For about four years, the ISEE-3 space probe stayed close to the* L_1 *Lagrangian point (upper left) in the earth–sun system. Its rocket was then fired to bring it back to the neighborhood of the earth. By interaction with the gravitational fields of the earth and the moon, the probe was sent out into space several times before finally, in December 1983, it was given such an impulse by the moon that it encountered the comet Giacobini–Zinner in September 1985.*

swung around the earth, nearly escaping again, only to return to and encounter the moon again on 22 December 1983. This time, they almost collided as the probe came within 100 kilometers from the surface of the moon! The strong gravitational attraction experienced at such a close pass to the moon gave such an impulse to the space probe that it was finally shot off into space, but not in any arbitrary direction. Just as a billiard player estimates how the ball will behave after the third or fourth rebound, the path of the probe had been worked out in advance: The moon would catapult ISEE-3 out into space so that it would pass through the tail of Comet

Giacobini–Zinner on 11 September 1985. The space probe and the cometary nucleus flew past one another at a distance of only 7800 kilometers.

This is a prime example of celestial billiards. ISEE-3 was brought back from the L_1 point to near the earth. The moon's gravitational attraction was used to send the probe on two excursions down the earth's solar-wind shadow and then to give it the right "kick" for an encounter with a comet that was moving on an elliptical orbit around the sun and was in the right place at the right time. With its new mission, the probe was given the new name *International Comet Explorer,* or *ICE.*

Herr Meyer as Simplicio

So far, we have considered the motion of bodies in gravitational fields in only a very simplified manner. We have assumed that a body in space obeys any gravitational force that acts upon it. The cannonball obeys the sun's attraction and moves in a circular, elliptical, parabolic, or hyperbolic orbit. The ISEE-3 space probe obeyed the gravitational fields of the sun, the earth, and the moon and allowed itself to be hurled out to where we wanted it to go. With an extended body, however, different points experience different forces, because the gravitational field of a celestial body is not equally strong everywhere. It declines with distance from the gravitating body. This has no effect on the motion of space probes, but the gravitational effect of the moon does differ at different places on earth and results in our experiencing tidal forces, which cause the ebb and flood of ocean tides, for example.

In order to illustrate the interplay of these forces, I shall now make use of one of Herr Meyer's dreams, as mentioned in the Foreword. Since his dreams are always prompted by his current reading, he often confuses and mixes different periods in history, as you will see in the following story.

Whenever Herr Meyer studied difficult subjects, he was only too glad to turn to lighter reading for relaxation. He was currently working through Galileo Galilei's *Dialogue Concerning the Two Chief World Systems,* and so, before going to sleep, he read a few pages of Jules Verne's *Journey to the Moon.* However, he was still unable to relax properly. In his mind's eye, he could still see the various speakers taking part in Galileo Galilei's dialog. When he finally closed his eyes, he found himself in a large palace.

He looked out of the window, but instead of looking onto a street, he saw a stretch of water between the palace and the house opposite. He saw a gondola coming from both the left and the right. Herr Meyer realized that he was in Venice. The gondolas brought two men in magnificent clothing to the portico. They came into the large room, in which Herr Meyer had previously been feeling a bit lost. The face of the taller man reminded him of Galileo as he had seen him in an old illustration. The two men came up to Herr Meyer and greeted him.

"Good day, Signore Simplicio," said the taller of the two. Now Herr Meyer realized that he was dealing with Salviati and Sagredo from Galileo's *Dialogue,* and that obviously they thought he was Simplicio, the fool. He was so surprised that he forgot to protest and to tell them his proper name. Later it seemed no longer convenient to call attention to the error, and so he let the matter rest. A conversation began immediately, which I shall attempt to repeat here word for word.

SALVIATI: "We agreed yesterday that today we would talk about falling bodies in space. Signor Galilei has proved that all bodies fall to the ground at the same speed. I was just thinking that this could have quite remarkable consequences. I imagined that I was inside a large box that had been dropped from the top of a very high tower and was falling toward the earth. I shall try to describe the actual state of falling rather than worry about the unpleasant consequences at the end of that fall. The box, any random contents in it, myself, and a few objects I had carried in with me, are all falling at the same speed toward the ground. If, while falling, I hold a stone and then open my hand to drop it, the stone will not fall out of my hand, because it is falling with me toward the ground. Since all bodies fall at the same speed, the stone will not move away from my hand. Inside a falling box, there seems to be no gravity, and the stone does not appear to be falling."

SAGREDO: "That seems to be because the box, you Salviati, and the stone are all falling at the same speed toward the ground."

MEYER: "If an astronaut lets go of something, it does not fall but remains apparently suspended in space."

SALVIATI: "Whatever is an astronaut? I have never heard the word before."

Before Herr Meyer could answer, he remembered that Jules Verne's *Journey to the Moon* did not describe a state of freedom from gravity.

MEYER: "Can you explain something to me that I have just read in one of Jules Verne's books?"

SAGREDO: "Who do you say? Jules Verne? I've never heard of that name. Who is he, a Greek philosopher? But the name doesn't sound Greek."

MEYER: "No, he is French and he wrote a story about several people who were shot toward the moon in a hollow cannonball. In the beginning, when the moon's attraction was still insignificant, the voyagers felt the gravitational attraction of the earth. But when they reached the moment when the forces of the earth and of the moon are equal, both the people and any objects they had with them floated around freely in their vehicle. They had entered gravity-free conditions, which lasted only a short time, however; then the attraction of the moon became predominant. The forces of the moon acted in the opposite direction: What was previously up inside the cannonball, became down and what was down became up."

SALVIATI: "That's a foolish story that your author, whose name I have never heard before and which I have already forgotten, has dreamed up. Let us ignore the fact that it will never be possible for men to travel to the moon; the events inside that hollow cannonball must have been completely different. You are quite right, dear Simplicio, to mention this ridiculous story while we are talking about freely falling bodies.

"As soon as the cannonball has left the barrel of the cannon, it is moving freely in space. It is not just that all falling bodies fall with the same speed, the same applies to all bodies that have been ejected with the same velocity. The ball, the men, and all the objects with them are flying upward at the same speed. Within the sphere, the earth's gravity is no longer detectable, and a stone that began with the same velocity in someone's hand is now moving upward with the same speed as the cannonball and its inhabitants. The stone will not fall out of its owner's hand. It is just the same as in my box falling from a high tower. When the cannonball finally reaches the region in space where the moon's attraction predominates, however, all the objects are still moving at the same rate. The shot, the travelers, and a stone in someone's open hand are moving at the same speed toward the moon. Thus, the stone still does not fall away from the hand, which is moving at the same speed as it is. The travelers will feel no gravity until they crash onto the surface of the moon. The story you just told us is obviously quite stupid."

Herr Meyer did not feel very happy. He had indeed sensed that something was wrong while he was reading *Journey to the Moon*, but he should have known from his school days what Salviati had just said. But before he could reply, Salviati continued.

SALVIATI: "I have realized that my explanation is not absolutely correct. Let's take another look at that inhabited cannonball that is falling onto the moon. I said that inside the cannonball the gravitational attraction from the earth's satellite would not be felt. That is not quite correct: A stone located on the side of the vehicle that is closer to the moon is itself closer to it and therefore feels a slightly greater attraction toward the moon than a stone on the opposite side of the cannonball's interior. The former is falling slightly faster toward the moon than the latter. The rigid body of the shot, however, is falling in accordance with the gravity acting at its center. Inside the freely falling cannonball, the stone closer to the moon appears to be slowly moving toward the moon, whereas the one farther away seems to move away from the moon, because the shot is falling slightly faster than it. Naturally, the movements of the two stones is very slow, since they are caused by the difference in the gravitational attraction of the moon at different points within the sphere."

While he was speaking, he took pencil and paper and drew the picture reproduced in Figure 21.

SALVIATI: "The travelers would feel no gravity at all in the center of their cannonball. In it, they can float around freely in space. On the other hand, on either side, the one toward the moon and the one away from it, the travelers will experience a small force that pulls them toward the walls."

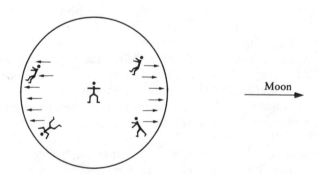

Figure 21. *Inside the hollow cannonball falling down onto the moon, weightlessness only reigns at its very center, where its center of gravity is located. On the side turned toward the moon, travelers would experience a small force directed toward the moon, whereas on the opposite side, the force would be directed away from the moon.*

SAGREDO: "Did you not present us with a similar argument yester-day when you explained the formation of the tides to us?"

SALVIATI: "It is actually somewhat similar. The earth and the moon both move around their common center of gravity. This point does not lie at the center of the earth but elsewhere, about a quarter of the earth's radius below the surface of the earth. On the side of the earth turned toward the moon, the moon's gravity is slightly stronger than it is on the side turned away from it, because its gravitational attraction decreases with distance." Salviati took the pencil and paper again and sketched the diagram shown in Figure 22. "On the side turned toward the moon, the gravitational attraction of the moon opposes that of the earth and thus raises a flood tide in the oceans. The earth is moving freely in the moon's gravitational field, just like my falling box. Therefore, at its center, the moon's gravitational force has no effect, just as the earth's gravity cannot be

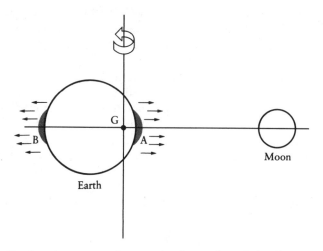

Figure 22. *During the course of a month, the earth and the moon move around their common center of gravity, G. At point A on the surface of the earth, which is closest to the moon, the earth's gravitational pull is lessened because of the opposing pull of the moon. The centrifugal force produced by the motion of A around G also tends to reduce the earth's gravity. Altogether then, the earth's attraction is reduced by two additional forces, causing the water to be heaped up in a tidal bulge. At the point farthest away from the moon, B, the moon adds to the earth's gravity, because it is pulling in the same direction. As point B is rotating around G, the water at that point experiences a very strong centrifugal force. The centrifugal force acting against the earth's gravity is so great that the gravitational attraction is decreased just as much as at A. Thus, a second tidal bulge occurs of approximately the same height.*

felt inside the falling box. But on the earth's side turned toward the moon, the latter's gravity is greater, and on the opposite side, it is smaller. On both sides, the water is raised up."

Herr Meyer was very confused by these words, because he remembered that Galileo had given a different explanation of the tides than this one, which was actually the same as Newton's and which is generally accepted nowadays as being correct. Herr Meyer had learned it in school.

MEYER: "Did not Newton explain the tides in a similar way, and didn't Kepler think along similar lines?"

SALVIATI: "Newton? Who's he? That's another name I have never heard before." (To himself: "So Kepler has been thinking about this as well? Perhaps I should have been reading his work more carefully in recent years.")

Herr Meyer realized that Salviati knew nothing about Newton since Galileo had died before Newton was born. So he put the matter aside and tried to envisage himself in Galileo's time.

MEYER: "If I have understood correctly, the same applies to the earth as it does to the cannonball falling down onto the moon. In the cannonball, the moon pulled the closer stone toward itself, while the other stone in the ball appeared to move away from it.
    "The same happens on the earth. The bodies of water closer to the moon are attracted more strongly toward the moon than the center of the earth. Therefore, they rise up and form a high tide. The bodies of water on the opposite side, like the second stone, are pushed away from the moon. That creates the second high tide."

Herr Meyer was indicating what he meant by pointing to Salviati's drawing (Figure 22).

MEYER: "The tides behave like this because the earth acts under the influence of the moon's gravitational attraction, just like the hollow cannonball. The earth is not just moving round the sun. Even a body as small as the moon causes it to swing around their common center of gravity once a month. And some fools in the highest authority maintain that the earth is the true center of the universe!"

Salviati and Sagredo looked at one another in astonishment. Signor Simplicio caught on very quickly today and was letting himself get carried away with statements, word of which had best not reach Rome.

When Herr Meyer woke up the next morning, he realized that in Galileo's time, he would probably have been reported to the Inquisition for the last statement he had made in his dream. They would have threatened him with torture and probably even shown him the instruments. Thus, when he found an official-looking letter from a Catholic agency in his mailbox, he instinctively felt a sudden queasiness. To his relief, it was only the Munich church contributions office reminding him to pay his contribution.

▪ The Catastrophe that Never Occurred

In Herr Meyer's dream, we saw that the gravitational attraction of the earth or of the moon can move a body in such a way that anyone on it no longer notices any gravity, as in Salviati's example of the falling box. But we also saw that for an object of any appreciable size that obeys gravitational forces as a whole, there are residual forces, known as *tidal forces*. We saw that these are responsible for the oceanic tides. They can also lead to sensational headlines, as was shown by the story of the "killer alignment" of 1982. "We are on the verge of a worldwide catastrophe," a German newspaper warned its readers on 25 July 1978. Further, the headlines proclaimed, "Volcanoes likely to erupt" and "Earthquake with 20,000 dead." Two archival photographs showed a town destroyed by an earthquake and a building about to be engulfed by a tidal wave. A third picture showed a "space professor," who explained that at the beginning of 1982, "tremendous gravitational forces" would cause dams to break. The year 1982 came and went without any special catastrophes. The rumors were dropped, and the professor turned to other spectacular issues.

Where did this announcement of catastrophe come from? One caption explained: "The nine planets are all in a line and will unleash devastating gravitational forces." Let us briefly discuss these "devastating gravitational forces" that the planetary alignment would exert on the earth. Since our planet reacts as a whole to the gravitational attraction of any celestial body or bodies, we only experience the tidal forces.

The moon continually exerts its tidal forces on the earth. As it rises and sets every day, the tidal forces act not only on the oceans alone, but also on the earth's atmosphere and on its entire body. The earth is being continuously massaged by the moon. The gravitational attraction of the sun also exerts tidal forces on the earth, and although they only amount to about half of those of the moon, we notice them particularly when they reinforce those of the moon and produce spring tides. Then the sun, the moon, and the earth are in a straight line. When these forces act against the moon's tidal forces, we have neap tides.

The planets also exert tidal forces on the earth. Venus produces the strongest effect when it is closest to the earth (at inferior conjunction; see Chapter 1). However, even then, its tidal forces amount to only one hundred thousandth of the moon's. Jupiter has the strongest gravitational attraction of all the planets, but even at its closest (at its opposition; see p. 56), it is so far away that the tidal forces it exerts on the earth are only one-tenth of those caused by Venus. We can forget the other planets; their tidal forces are far weaker.

Thus, if the sun and the moon fail to produce a particular effect on the earth, then the other planets will certainly not, even when they all line up for a tug-of-war. And at the time of the "catastrophe" in 1982, they were really not at all in a line but actually quite scattered.

The tidal forces of the aligned planets on the sun, which were also dragged into the headlines to fill the "silly season" with sensational forecasts, are quite insignificant. In fact, the sun did not appear disturbed in 1982 and its effect on the earth was no different than in other years.

The "Jupiter effect" was just a newspaper canard, comparable to the anxiety over the possible collapse of a perfectly solid dam. Think of a dam that for decades has withstood the pressure of the water confined behind it: Within it the pressure forces caused by the water and the tension in the concrete wall are in balance. Occasionally, a sparrow sits on top of the wall. That hardly upsets the balance between the forces exerted by the water and the concrete; so the dam doesn't break. The fuss over the Jupiter effect was just as if someone had predicted that the dam would break if nine sparrows decided to perch on the wall at the same time.

THREE

Planet Earth

When Švejk later on described life in the lunatic asylum, he did so in exceptionally eulogistic terms: ". . . One [person] was in a straitjacket all the time so that he shouldn't be able to calculate when the world would come to an end. . . . And I also met a certain number of professors there. One . . . explained to me that inside the globe there was another globe much bigger than the outer one."

JAROSLAV HAŠEK (1883–1923)
The Good Soldier Švejk [Trans. Cecil Parrott]

■

We can certainly say a lot about our own planet, because we know more about it than about any other celestial body. But I shall try to discuss the earth in the same way as the other planets in the solar system, particularly with regard to its interaction with the sun and the moon.

■ *An Ellipse—Almost a Circle*

The planet earth moves in an elliptical orbit around the sun, as dictated by Kepler's first law. The sun is not at the center of the ellipse but at one of the two foci. At certain times of the year, the earth is closer to the Sun than it is at others. But the elliptical orbit is nearly a circle, and although at the beginning of the year we are five million kilometers closer to the sun than in July, the difference is only a small fraction of the average distance between the earth and the sun, which is 150 million kilometers. Light, traveling through space at the inconceivable velocity of 300,000 kilometers per second, takes eight minutes to cross that distance. Only when we use sophisticated measuring instruments do we discover that our distance from the sun varies slightly during the course of one orbit, that is, over a year. The difference between summer and winter seasons has nothing to do with the changing distance between the sun and the earth. On the contrary: In the northern hemisphere's winter, we are slightly nearer the sun than in summer.

As the earth revolves around the sun, it also rotates. During one orbit, it rotates 365¼ times on its own axis. The revolution around the sun causes the seasons, and the rotation produces the alternation of day and night. A year, therefore, amounts to 365 days and 6 hours. After four orbits, the extra 6 hours that remain at the end of each year have added up to a whole day; we adjust for this every four years by adding an extra day, a leap day, to February.

The earth's rotational axis is not at a right angle to the plane of its elliptical orbit. This produces our summer and winter. Figure 23 shows how the seasons arise in the course of an orbit. The diagram also shows why the seasons in the northern hemisphere are shifted by six months when compared with those in the southern hemisphere. When we have our summer in the north, it is winter in the south, and vice versa.[1]

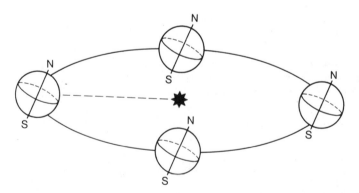

Figure 23. *The causes of the seasons on earth. The earth's axis is not perpendicular to the orbital plane. Once a day the earth rotates around this fixed axis. On the left, more of the sun's radiation is falling on the northern hemisphere: It is summer. On the right, more sunlight is falling on the southern hemisphere: It is winter in the northern hemisphere.*

Since, according to Kepler's second law, a planet moves faster in its orbit when it is closer to the sun, winter in the northern hemisphere (i.e., summer in the southern hemisphere) is somewhat shorter than the northern summer (and the southern winter).

It was recognized quite early that the earth is round, but its size was determined relatively late. To the very end of his life, Columbus thought that he had reached Japan or India, because in his day, the size of the earth had been underestimated. Yet Eratosthenes of Cyrene (276–195 B.C.) had explained the principle by which the diameter of the earth could be determined. Using this method, shown in Figure 24, the Frenchman Jean Picard (1620–1682) determined the size of the earth in 1671 by using the distance between Paris and Amiens. When Newton learned about this measurement, he was able to use it in his new theory of gravitation. The earth's radius at the

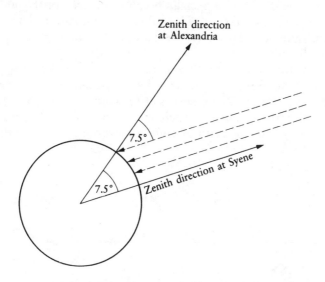

Figure 24. *How Eratosthenes determined the size of the earth. At the summer solstice, the sun is at the zenith at Syene. At the same time, to the north, at Alexandria, it stands about 7.5 degrees south of the zenith. From this, Eratosthenes concluded that the included angle between the two lines connecting the center of the earth with Syene and Alexandria must be 7.5 degrees. (For clarity, this angle is greatly exaggerated in the diagram.) Thus, the fraction represented by the distance between Syene and Alexandria divided by the whole of the earth's circumference must amount to 7.5/360. Because this distance was known, the earth's circumference and thus the earth's radius could be calculated. We cannot be certain whether Eratosthenes obtained the correct result. Later, the same procedure was used to determine the earth's radius, but instead of the distance between Syene and Alexandria, a precisely measured distance was used.*

equator amounts to 6378 kilometers. As we shall see later, the earth is flattened, and its radius at the poles is somewhat smaller.

Knowledge of the size of the earth allows us to determine distances in the solar system. The principle is shown in Figure 25. As the earth, carrying us with it, rotates around its axis, a body out in space appears in slightly different directions at different times of the day. When we compare it against the more distant backdrop of the fixed stars, it appears to move. The closer it is, the faster this apparent movement. If we know the size of the earth, then we also know the radius of the circle around the earth's axis that we trace out in a day during the course of one rotation. Take the sun as an example. In addition to its motion across the sky during the course of a year, it should also show a daily back and forth movement, since we observe it first from one side of the earth (at sunrise) and then from the other (at sunset). This shift of the sun amounts to only a few seconds of arc, which is only about one one-hundredth of its diameter. Such a small angular shift in the sun's motion cannot be detected, especially because the light of the sun drowns out the light of all the fixed stars that could provide suitable reference points for such a measurement. However, there are ways to determine the distance of the sun indirectly.

As it revolves around the sun, Mars comes close to the earth, namely when it appears on the opposite side of the sky to the sun as seen from earth. Astronomers then say that the planet is at opposition (see also p. 122). The martian orbit departs relatively strongly from a circle and, therefore, comes considerably closer to the earth at some oppositions than at others. The

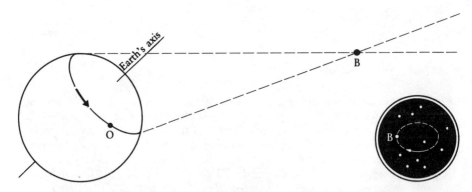

Figure 25. *During the course of a day, an observer O sees a celestial body B in different directions. If it is observed in front of the very distant, fixed stars, the object appears to describe a small ellipse during the course of a day. The sketch at lower right shows this apparent motion of the body as viewed through a telescope. This phenomenon is known as* diurnal parallax.

oppositions at which it is particularly close occur at 15-year intervals. If, at that time, it is observed simultaneously from two points on earth, its shift relative to the fixed stars can be determined. In 1672, French astronomers in Paris, working under Jean Richer (1630–1696), measured the position of Mars when it was at opposition. At the same time, it was also observed by an expedition that had been sent to Cayenne. The distance between Paris and Cayenne was known — Jean Picard had determined the diameter of the earth the year before. From the slightly different positions of the planet against the background stars as observed at the two points, it was possible to estimate the distance between Earth and Mars on that date as being 10,800 times the earth's diameter. But people wanted to know the distance from the earth to the sun, not to Mars. Kepler's third law provided the answer. The earlier measurements had provided the difference between the radii of the orbits of Earth and Mars. (For the sake of simplicity, we will assume here that we are dealing with circular orbits. The computations are slightly more complicated for elliptical orbits, but the principle remains the same.) Since the orbital periods of Earth and Mars were known (which are one year for the earth and one year and 321.98 days for Mars), the ratio of the orbital radii could be calculated from Kepler's third law. From the measured difference and the ratio between the two orbital radii, the two radii themselves could be calculated,[2] and thus the distance of the sun was first determined. The expedition to Cayenne at the time of the opposition of Mars yielded another very important result. It was noticed that the earth was not truly spherical but slightly flattened.

The English astronomer Edmond Halley (1656–1742) had already suggested a completely different method of determining the distance between the earth and the sun. From time to time, the planet Venus appears to cross the sun's disk as seen from Earth, in a *transit of Venus*. This does not happen very often, but it occurred in 1631, 1639, 1761, 1769, 1874, and 1882. Since we see the unilluminated side of Venus, we observe its transit as a small black dot passing across the brilliant face of the sun. The first transit of Venus of this series had already been predicted by Kepler. However, the event was not observed even from those locations on the globe where it occurred during daytime hours. The next transits of Venus that we shall be able to see will occur in 2004 and 2012. During a transit, Venus is closer to the earth than Mars is at a favorable opposition. When observed from different locations, Venus appears to cross the disk of the sun at different positions. From this, the Earth–Venus distance can be determined, and therefore the difference between their orbital radii. By again using Kepler's third law, we find the ratio of the orbital radii; and from the sum and the ratio between the radii, we can determine the individual orbital radii.

The next transit of Venus after Halley had suggested this method came in 1761, 19 years after Halley's death. Numerous observers in various parts of the world hoped to observe the dark spot passing across the disk of the sun.

However, when Venus appeared against the sun, a hitherto unknown optical effect was noticed that prevented accurate measurements. At this point, I do not intend to discuss this "black-drop phenomenon" any further, but it led to less accurate results than those obtained when the French observed Mars in 1672.

When Captain Cook sailed around the world in his ship HMS *Endeavour* between 1768 and 1771, he stopped in Tahiti, at a place still known as "Point Venus," in order to observe the transit of Venus in 1769. Cook's measurements were not analyzed until many decades later by the German astronomer Johann Franz Encke (1791–1865) and did not turn out to be particularly accurate.

A total of 50 expeditions were mounted for Venus's transit of 1874, all with very meager results. A relic of the German undertaking, carried out at the astronomical cost of 600,000 gold marks, can still be seen at the Bamberg Observatory in the form of the expedition's heliograph, with which Venus was photographed in front of the sun. After 1874, interest in transits of Venus declined. A better method of determining the Earth–Sun distance had been found.

Exactly since the first day of the nineteenth century, it has been known that in addition to the major planets, there are innumerable small bodies that revolve around the sun in orbits between those of Mars and Jupiter. These bodies will be discussed in detail in Chapter 8. Many of their orbits deviate considerably from circles. In particular, two of these minor planets occasionally come very close to the earth. They are called Eros and Amor. By observing them from several places on the earth, astronomers can determine their positions on the sphere of fixed stars very accurately, far better than for Mars. This gives the difference in the orbital radii of the earth and the minor planets. Kepler's third law then again gives the ratio of the orbits from their orbital periods. Finally, we have yet another Jack-and-his-dog problem to solve (see Note 2 to this chapter). Because the minor planets come closer to the earth than Mars, they provide more accurate values for the distance between the sun and the earth. Thus, the minor planets Eros and Amor allow astronomers to choose either an "erotic" or an "amorous" method of determining the earth–sun distance.

More recently, yet another method has been used. A radar pulse is sent toward Venus, and the time between the emission of the pulse and the return of its echo is measured. From the time taken by the signal, which travels at the speed of light, the Venus–Earth distance can be determined, and thus the difference in the orbital radii. Kepler's third law again gives us the ratio of the radii, from which we again obtain the individual orbital radii. Once we have the radius of the earth's orbit, we can use the same law to find the orbital radii of all the planets from their known orbital periods. In September 1958, it even proved possible to obtain radar echoes from the surface of the sun itself. The solar system has therefore been thoroughly surveyed.

▪ *The Rotating Sphere*

The earth rotates, and as we stand on its surface, the fixed stars seem to be carried on a sphere that rotates around us. The earth's rotational axis indicates the two points around which the heavens appear to rotate, the north and south celestial poles.

The rotational velocity is not so great that centrifugal force flings us off into space, but even though it acts on all bodies that rotate with the earth, it is very small. It is strongest at the equator, where the centrifugal force acts vertically upwards on every object, and therefore makes it lighter. Thus, everything weighs about five parts per thousand less at the equator than at the poles. As a result of its rotation, the earth is not a true sphere; centrifugal force flattens it slightly. We have already seen that the deviation of the earth from a sphere was first noticed by the French expedition to Cayenne in 1672. The flattening is not very significant, however: The equatorial radius is only 21 kilometers greater than the distance between a pole and center of the earth. The earth is girded with an equatorial bulge.

Even without looking out into space, it is quite possible to discover that the earth rotates. The rotation of our planet is perceptible even in a closed, windowless room. In quite a few institutions, such as the National Museum of American History in Washington, D.C., the United Nations in New York, and the California Academy of Sciences in San Francisco, the rotation of the earth is impressively demonstrated by a Foucault pendulum, a giant pendulum swinging in a multistory space.[3] In Washington, a 108-kilogram weight hangs at the end of a 16-meter-long wire and, due to the length of its wire, swings very slowly. It takes the pendulum nearly 5 seconds to go from one side to the other (and the same time to return, of course). From the motion of the pendulum visitors can see with their unaided eyes that the earth rotates. For simplicity, we shall assume that the pendulum is hanging over the north pole. It is set in motion, and tends to remain swinging in a single plane. Figure 26 shows the pendulum and the plane in which it swings. It continues to swing in this plane while the earth rotates about its axis. So the pendulum is not affected by the rotation of the earth, because the earth's gravitational attraction is the same whether it is rotating or at rest. But to someone standing on the earth and rotating with it, it will appear that he is standing still and that it is the plane in which the pendulum is swinging that rotates. The rotation of this plane is a sign of the rotation of the earth.

With a pendulum swinging at the north pole, the situation is very simple, because its suspension point lies directly above the earth's rotational axis and therefore remains stationary. But museums are not usually located at the north pole, and therefore the situation is more complicated, because the suspension point is turning in space with the earth. Despite this, an observer can still see, after a few swings of the pendulum, that the plane in which the pendulum is swinging is slowly rotating relative to the surface of the earth.

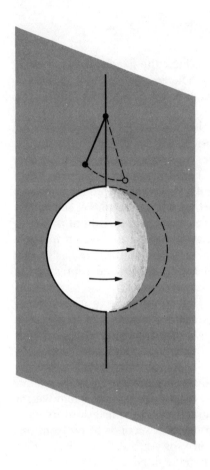

Figure 26. *A pendulum suspended over the north pole swings in a specific plane (shown shaded). The hemisphere in front of this plane is drawn with a continuous line, whereas that behind the plane is shown by the broken line. The plane of oscillation does not take part in the earth's rotation. As the earth turns, it appears to an observer that it is stationary and that the plane of oscillation is rotating.*

The experiment is quite unconnected with astronomical observations. Even if the earth were completely covered with clouds, and the sun and stars were never seen, the rotation of the plane of Foucault's pendulum would still show that we live on a rotating earth.

Although the rotation of the earth can be easily demonstrated by a Foucault pendulum, the first physical experiment that showed that we live on a rotating body had already been carried out decades earlier. Since the time of Newton, it had been known that freely falling bodies would behave differently on a rotating body than on one at rest. Many attempts were made to prove this, but results of the measurements were ambiguous. A conclusive experiment was carried out from the tower of the Church of St. Michael in Hamburg by Johann Friedrich Benzenberg (1777–1846), who had studied theology before turning to science. As a student in Göttingen, he had already made the most important astronomical discovery of his life, which we shall discuss in greater detail in Chapter 8. In 1802, two years after obtaining his doctorate, he investigated the fall of small lead spheres from the top of the church tower. The rotation of the earth was supposed to affect the spheres so that they would not fall perfectly vertically following the direction of a plumb line.

This seemingly strange behavior of a falling body is linked to an important concept in mechanics: the conservation of angular momentum. In order to understand it, we shall carry out a small thought experiment.

Tie an object, such as a spoon, to a piece of string and then, holding it above your head with one hand, start to swing it round. The object will be circling around your head, just as a planet goes around the sun. Once it has been set in motion, it continues to move at a constant speed in a circular orbit. Only over a period of time will air resistance slow it down, unless you keep it in motion with small movements of your hand. If there were no air resistance, the object on the end of the string would continue to revolve around your head with the same velocity forever. Now imagine using your free hand to pull a little bit of the string from the other hand, thereby shortening the string tied to the object. This will cause the object to move faster, because with the radius of its orbit having been decreased, its revolutions around your head now take less time. Its velocity has increased, without your having had to give it any additional impetus.

With a stopwatch and a ruler, it would not take long to work out the following rule: half the radius, twice the velocity. As halving the radius means halving the circumference of the circle, the object now moves with twice the velocity around a path that is only half the size. Thus, it takes only a quarter of the time to complete one revolution. In other words, multiplying the velocity of the body by the length of the string gives the same value in both cases. However you alter the length of the string—as long as you do not slow down the body or give it any additional impetus—the product of the velocity times the radius remains the same. In physics, this is known as a *conserved quantity*. With our revolving object, the product of velocity and radius gives us the *angular momentum* of the body. Even though you change the radius of the orbit, its angular momentum is conserved.

We are all familiar with the conservation of angular momentum from watching an ice skater doing a pirouette. When she draws in her arms, she rotates faster, because of her angular momentum. Every atom in her body is rotating about her axis. Her angular momentum consists of the sum of the moments of all her atoms. If we multiplied the velocity of each atom by its radius from the skater's axis and took the sum of the results, we would obtain the skater's angular momentum. (Luckily, we do not really have to carry out this complicated calculation for all the atoms, because there are easier ways of determining the angular momentum of any object.) When the skater brings her arms in toward her body, the distances between the atoms in her arms and the axis of rotation become smaller. If her rate of rotation stayed the same, we would obtain a smaller value if we were to carry out the same calculation again. But her angular momentum cannot alter. If the distance from the axis becomes smaller, then the speed of rotation must increase for the angular momentum to remain the same: The skater rotates faster.

We are already familiar with another example in which the angular momentum plays an important part, namely Kepler's second law (see Chapter 1). We saw that a planet travels faster along those parts of its orbit that are closer to the sun. This is because of the conservation of angular momentum. As a planet moves along its orbit, its angular momentum must always remain the same. Multiplying its orbital speed by its distance from the sun must give the same value when it is closest to the sun as when it is most distant. Therefore, it follows that its velocity must be high when its distance from the sun is small.

But back to the church-tower experiment. Just as it was easier to understand the Foucault pendulum when we imagined it as hanging over the north pole, so it is easier to understand the behavior of a falling body if we imagine that the tower is on the equator (Figure 27). Let's assume that a man at the top of the tower holds a sphere that he suddenly releases, so that it can drop freely. Initially, the tower, the man, and the sphere are rotating with the rest of the earth around the earth's axis. The sphere is therefore moving in a circle with a radius equal to that of the earth plus the height of the tower. The earth's rotation carries the sphere in the man's hand from west to east. When he lets go of the sphere, it does not merely fall downward. It must retain its angular momentum about the earth's axis. That means that as it falls, the product of its rotational velocity and its distance from the earth's axis must remain the same. As it reduces its distance from the center of the earth by

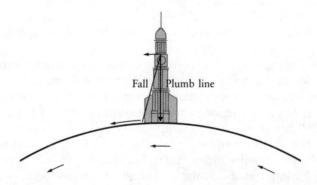

Figure 27. *A sphere dropped from a tower on the equator deviates from the direction indicated by a plumb line. The effect, shown here for the equator, is also noticeable at other latitudes, with the exception of the poles, even though the motion of the body is more complicated elsewhere than it is at the equator.*

falling, its velocity must increase. During its fall, it therefore moves faster from west to east than the tower, the observer, and the ground it reaches. It hits the ground farther to the east than the point indicated by a plumb line. Benzenberg's measurements using the tower in Hamburg hinted that the path of the falling sphere indeed deviates from the plumb line. Two years later, he repeated the experiment in a coal mine near Dortmund. His results indicated the rotation of the earth, but the first definite proof came with the experiment that Ferdinand Reich (1799–1883) carried out in a mine shaft in Saxony.

It is very difficult to indicate the rotation of the earth by carrying out experiments with falling bodies. The sphere must be released very carefully, without giving it the slightest sideways motion. This is nearly impossible to avoid unless the sphere is first suspended from a thread that is then very carefully severed. Air currents not only affect the sphere, but they may also set the plumb line (required for the purpose of comparison) swinging. This is why the Foucault pendulum is used exclusively nowadays as a simple means of showing the rotation of the earth, because it does not require any such complicated arrangement.

▪ *The Wobbling Earth*

Life would be much simpler if there were only two bodies, namely the earth and the sun, and if all other bodies were so far away that they did not disturb this idyllic state. The existence of the moon makes life much more complicated for astronomers. It is the third body in the sun–earth–moon system. We must remember that the earth is slightly flattened because of its rotation. The sun and moon exert a greater force on the part of the equatorial bulge that is nearer to them than they do on the part that is turned away from them. As indicated in Figure 28, the moon tries to pull the earth's axis into a position perpendicular to the plane of the moon's orbit. The sun acts in the same way, trying to force the axis to lie perpendicular to the earth's orbit around the sun. How does the rotating body of the earth react to these forces?

Consider a spinning top (Figure 29) that is not precisely perpendicular to the ground. If it were not rotating, it would topple over because the earth's gravity is acting to bring its axis as close to the horizontal as possible. But the situation is completely different if the top is rotating. In that case, it will not topple over as a result of the earth's gravitational attraction but instead, it will start to wobble. And just like the toy top, the earth also reacts by a wobbling motion to the sun's and the moon's attempts to pull it upright. The

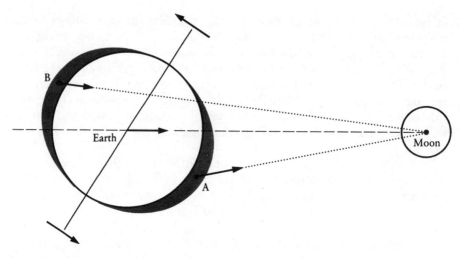

Figure 28. *The gravitational attraction of the moon on the flattened earth acts more strongly on the equatorial bulge on the closer side A than it does on side B. The moon therefore tries to "straighten up" the earth's axis, which is inclined to the moon's orbit, as indicated by the arrows at the poles. But the rotating earth reacts like a spinning top (see Figure 29).*

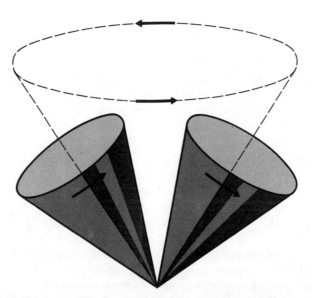

Figure 29. *The downwardly directed gravitational attraction of the earth tries to upset a top. If the top is not rotating, it does fall over. But when it is rotating, it does not react directly to the force of gravity, but instead, its axis describes the surface of a cone.*

earth's axis revolves in space along the surface of a cone (Figure 30). Although today the earth's axis is pointing very close to the pole star, it did not always do so. During the period of the ancient Egyptians, our present pole star was far from the celestial pole. At that time, about 3000 years ago, the celestial sphere rotated around a star in the constellation of Draco. In another 12,000 years, Vega, in Lyra, will be the pole star.

The Earth's wobbling motion caused by the sun and the moon has a period of about 26,000 years. After that period of time, our current pole star will again guide nighttime hikers on their way—provided there will still be hikers. This wobbling motion, called *precession* by astronomers, was known to the ancient Greeks.

Besides causing this slow gyration of the earth's axis in space with a period of many thousands of years, the sun and the moon are also responsible for a far more obvious phenomenon. The sea rises and falls twice a day with the flood and ebb tides. From one day, to the next, the flood tide begins about 50 minutes later. The moon also sets 50 minutes later each day. This correspondence must have been noticed in very early times. The explanation of this phenomenon came in several steps. Kepler recognized the true nature of the problem, whereas Galileo gave an incorrect explanation; but Newton explained the tides in the same way as we understand them today and as they were described in Herr Meyer's dream in Chapter 2.

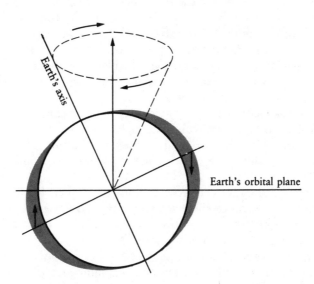

Figure 30. *How the earth's axis reacts like a top to the forces of the sun and moon trying to pull it upright. The axis describes a cone, so that in 26,000 years, the north celestial pole traces out a complete circle. At present, the north celestial pole is close to Polaris, the pole star.*

▪ *The Rotation of the Earth and the Rhythm of the Tides*

The moon produces two tidal bulges in the oceans. Because of the rotation of the earth, these bulges seem to sweep around the earth from east to west. In addition, the sun causes smaller bulges. We notice its effects particularly when the sun and the moon are in a straight line, namely at full or new moon. (We already mentioned spring tides in Chapter 2.)

When the Swabian doctor and physicist Julius Robert Mayer (1814–1878) discovered during a sea trip one of the most important laws of physics, the law of the conservation of energy, he also asked himself what sustained the ebb and flow of the tides. He had no idea that about 100 years earlier, Immanuel Kant (1724–1804) had also pondered this problem. We shall describe their ideas in the terms used in modern physics. For the sake of simplicity, we shall ignore the influence of the sun on the tides and just consider the tidal effects caused by the moon.

When the Moon raises the tidal bulges in the seas while the earth simultaneously rotates beneath them, an enormous amount of energy must be expended. For example, the flood tide fills bays and estuaries, which are then emptied several hours later. If a dam is erected across an estuary, the back-and-forth motion of the water can drive a turbine and produce electricity. Such a tidal power station currently exists in France at the mouth of the river Rance near Saint-Malo, where the difference between high and low water is about 10 meters. At the beginning of the flood tide, the seawater used to flow freely into the mouth of the river, but now it has to squeeze through narrow sluices in the 750-meter-long wall of a dam, driving turbines as it does so. The water again passes through the blades of the turbines when the water recedes. This power station produces 240 megawatts, which is not a lot. Moreover, the power is not available around the clock; the system operates at full capacity only four times a day during the flood and ebb of the two tidal bulges. It seems unlikely that such power stations can solve mankind's energy problems. But where does the energy produced by this French power station actually come from? What is the source of energy that men have tapped?

The answer had already been given by Kant and Robert Mayer: The energy comes from the rotation of the earth. Anyone who has ever tried to stop a spinning flywheel knows that rotation stores energy. If one tries to stop the wheel by friction, such as by pressing one's hand against the wheel as it is spinning, one will realize that the energy of rotation is converted into heat. If one lifts up the wheel of a bicycle and sets it spinning rapidly and then lets it turn a dynamo, the lamp burns until the energy in the rotating wheel has been exhausted, with part of the energy being turned into electrical energy. The rotation of the earth is braked by the tides, not just because of the laughable 240 megawatts from the Rance power station, but because the tides themselves consume and dissipate energy.

Let us first consider a simpler example, assuming that the earth rotates so slowly that it takes 27 days to complete one rotation, the same time it takes the moon to complete one orbit relative to the fixed stars. In that case, the earth would always turn the same face toward the moon. For the inhabitants of that side of the earth, the moon would always hang motionless in the sky and would neither rise nor set, and the two tidal bulges would be stationary on the earth. One of them would be centered on the side that faces the moon, and the other on the opposite side; there would be no tides. The state of high or low water on a particular stretch of coast would remain the same forever. In reality, of course, the earth rotates much faster. A tidal bulge that is in the center of the Atlantic at a specific time travels toward the American continent. We could just as well say that, because of the earth's rotation, the continent is moving toward the bulge. Complicated currents must occur as America nears the tidal bulge. The earth's rotational energy is converted into motion of the water masses. Omnipresent friction then converts the energy in the motion into heat. The two tidal bulges exert a braking force on the earth, just like the pair of brake shoes inside the brake drum of a vehicle. The day is therefore gradually increasing in length, even though by only about two thousandths of a second per century.

The English astronomer Edmond Halley was the first to suspect that there was something odd about the earth's rotation when he tried to reconcile old observations of eclipses with observations made in his day. Although no exact times were given in the old observations of eclipses, the places where they were observed are well established. When Halley calculated backwards, he found that eclipses occurred at places where according to his calculations, no such event should have been visible. Did the moon revolve around the earth somewhat slower in those days? We now know that the earth was actually rotating somewhat faster. The earth, which previously served as a clock, had slowed down. If my watch runs slower without my noticing it, then events in the world around me appear to occur faster. If my watch runs at half the normal rate, trams that usually run six times an hour would appear to be arriving twelve times an hour. If anyone were to set a clock by the earth's rotation, in the belief that it does not alter, any regular processes occurring in the universe, such as the revolution of the moon around the earth, would appear to become faster with time. Two thousandth of a second per century seems to be an imperceptible amount. But over a long span of time it adds up!

Two thousand years ago, the day was, on average, indeed about two hundredth of a second shorter than it is now. In the 730,000 days that have passed since then, the errors have accumulated to total $0.02 \times 730,000 = 14,600$ seconds. That makes four hours, a definite effect that can considerably influence the occurrence of eclipses. If the earth's clock is four hours different from the moon's, total solar eclipses will be seen at different points on the earth's surface than one would otherwise naively expect. It is therefore not surprising that Halley came across old eclipses that occurred at places where they could not have been visible if the earth's rotation had been constant.

▪ *The Moon is Moving Farther Away*

The tides do not only brake the earth's rotation; they also influence the moon. The tidal bulge is not actually highest at the exact point on the earth's surface that is closest to the moon, namely where the moon is at the zenith. Because of friction, high water does not arrive until about $2^1/_2$ hours after the moon is at its highest, and the high water on the opposite side of the earth is similarly delayed. This delay influences the moon's motion.

Consider the tidal bulges from the point of view of the moon. They do not lie on the line joining the centers of the earth and the moon. Because the earth rotates faster than the moon moves in its orbit, the two tidal bulges seem to be carried slightly forward by the earth's rotation. Seen from the moon, they appear to lie ahead of the earth–moon line. But it is not just the solid body of the earth that contributes to the gravitational force on the moon; the two tidal bulges exert an additional attraction (Figure 31). The moon is therefore not only subject to the gravitational pull acting through the moon's center of mass, but to an additional acceleration caused by the tidal bulges, as indicated in the diagram. As a result, the centrifugal force acting on the moon is continuously increasing. It is slowly being forced outward, and the radius of its orbit is increasing. The effect is not great: The moon recedes by about 4 centimeters per year, a vanishingly small amount when compared to its distance of 384,000 kilometers. If this has always been the case, billions of years ago the moon must have had a much smaller orbit around the earth, and the tides must have been correspondingly stronger. The earth must have experienced much wider tidal ranges, and the friction may even have been so great that the oceans were significantly heated. However, nothing indicates that the earth has been subjected to such a tidal catastrophe in the last 1 to 2 billion years. Friction in the oceans may have been smaller in the past. Could the moon have always been at a fair distance, and is it only recently being pushed further from the earth by the earth's tidal bulges?

▪ *The Equable Planet*

As the earth rotates, its day side captures a fraction of the radiation that the sun emits in all directions. The amount intercepted is not great, being less than a millionth of the total radiation. But this tiny fraction makes life on

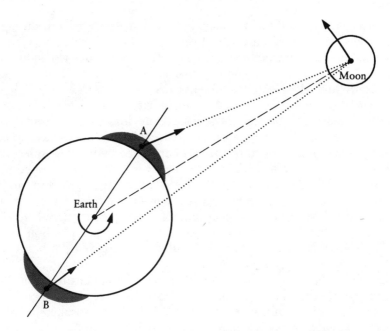

Figure 31. *The earth and moon are shown here as if we were looking down on the north pole. The moon, its distance, and the tidal bulges are not drawn to scale. The moon produces the oceanic tidal bulges. Since the earth rotates faster than the moon orbits around it, it effectively rotates beneath the tidal bulges. Because of friction, it tends to cause the tidal bulges to advance slightly. Therefore, they lie ahead of the moon and tend to accelerate it. As can be seen in the illustration, the line joining the centers of the two tidal bulges does not point toward the moon but toward a point on its orbit that the moon will only reach somewhat later. This affects the motion of the moon. Because the gravitational attraction of the tidal bulge on the side closest to it is greater than that of the bulge on the other side of the earth, the moon experiences a slight acceleration in the direction of its motion along its orbit.*

earth possible. Part of the radiation falling on the day side is reflected back into space. The rest warms the earth's surface and the lowest layers of the atmosphere, creating bearable temperatures for us. Solar radiation comes from a hot body, the temperature of the radiating surface being around 5500°C. At that temperature, a body radiates primarily visible light. This is not an accident. Our eyes have evolved so that they are most sensitive to wavelengths where light falling on earth is strongest. Our planet is heated by the incident radiation, but a portion of the energy is radiated back into space. This radiation, known as *infrared radiation,* is invisible to us. The balance between incident and emitted radiation determines the temperatures on our planet. We should be thankful that the earth rotates relatively rapidly, because it absorbs far too much energy by day. Only the emission during the night makes life possible on earth. If the earth always turned the same side toward the sun, the temperature throughout this endless day would probably be 100°C hotter than it is now, while all the Earth's water would be permanently locked in ice on the night-side.

We can also thank the atmosphere for our tolerable climate, because it allows visible sunlight to reach the surface. Only during heavy cloud cover is a large part of the incoming radiation reflected back into space. But the outgoing radiation is even more strongly controlled by the atmosphere than incident radiation. The less transparent the air is to invisible infrared radiation, the more difficult it is for the earth to radiate away any heat, and the hotter it becomes. Carbon dioxide, which is produced by the combustion of organic materials such as wood, coal, and oil, is opaque to infrared radiation. This is why it may be dangerous for us to pump too much carbon dioxide into the atmosphere. By the time all the oil and coal have been burned, the climate will have changed considerably.

It appears to be a very lucky chance that the earth has an atmosphere that helps to regulate the climate, because the existence of a gaseous shell around any planetary body cannot be taken for granted. The atmosphere is held by the earth's gravity, but it is not particularly tightly bound.

If one throws a stone upwards, it generally comes straight back. If one were to give it an initial velocity of 11.2 kilometers per second, then the earth's gravitational attraction would no longer be able to retain it, and it would escape into space. The earth is unable to retain bodies that are moving too rapidly away from it. What we experience as the heat of a body is no more than the motion of its constituent atoms and molecules. In a hot piece of iron, the atoms are rapidly vibrating back and forth. In the atmosphere, the molecules of the various gases are colliding with one another at high velocities. The higher the temperature, the higher the velocities of the various sorts of molecules. The lightest molecules move fastest. In air at room temperature, say around 19°C, oxygen atoms are moving with an average speed of 675 meters per second. The atoms of the much lighter hydrogen, on the other hand, move about four times as fast. But their velocity is still far below escape velocity. Despite this, we cannot expect the earth to retain all the molecules in the air forever. The values just given for air at room temperature are only average values. There are always a few atoms that are moving considerably faster, just as there are always some that are moving much slower. In the outer layers of the atmosphere, the loss of hydrogen is particularly significant. The heavier atoms, though, appear to be retained by the earth—at least the loss rates are so low that over its lifetime, the earth can have lost only very small quantities of those gases. The cooler a planet is, the easier it is for it to retain its atmosphere, because then the molecules move more slowly. It is also advantageous for the planet to be massive and of small diameter, because then the gravitational force at its surface is strong and the escape velocity high.

▪ The Magnetic Earth

Shortly before midnight on 27 August 1958, an experiment codenamed *Argus* was carried out high over the Atlantic. Initially, a bright flash was seen at a single point in the sky, which then spread north and south, glowing like a polar aurora. It was as if the aurora borealis had come to the Azores. The *Explorer IV* satellite, which was in orbit at that time, transmitted data to its ground station indicating that the earth and its atmosphere had been surrounded by a shell of electrons. This experiment was planned and prepared by the United States in response to *Sputnik,* the first artificial satellite, launched by the Soviet Union the year before. The United States needed a success. During the *Argus* experiment that night, and subsequently, a total of three atom bombs were exploded outside the earth's atmosphere enabling scientists to study the diffusion of the electrons emitted by the radioactive fission products of the explosions, in the earth's magnetic field.

Since the second century A.D., when the Chinese discovered the compass, it has been known that the earth has a magnetic field. We are unable to see it and do not even notice the particles trapped within it directly. But magnetometers reveal invisible structures surrounding the earth, and particle counters give us information about the charged particles that are trapped within them. It is a very diverse world, but we have developed no senses to deal with it; we can only recognize it through our instruments. Columbus realized on his very first voyage that a compass needle does not point due north, because although the deviation of a magnetized needle from true north is extremely small in Europe, it becomes much greater the farther one sails out into the Atlantic.

There are various theories as to why the earth has a magnetic field. We shall see later that other planets are also gigantic magnets. But the basic cause of the earth's magnetism is still unknown. Just as current is produced in a bicycle dynamo by the motion of electrical conductors, currents probably arise within the earth through the movement of electrically conducting, fluid material. These electrical currents themselves produce a magnetic field, in just the same way as a coil of wire, through which a current is flowing, becomes magnetized. If this is correct, celestial bodies that do not have fluid cores will not be magnetized. This is probably the reason why the moon has no magnetic field.

The earth's magnetic field reaches far out into space where it encounters the solar wind (see p. 151). This is why the earth has a long magnetic tail, stretching out into space on the side away from the sun. (This was mentioned when we discussed the path of the *ISEE-3* space probe in Chapter 2.) Most of the particles in the solar wind are hydrogen atoms, or rather individual nuclei and electrons. Both the nuclei (protons) and the electrons are electrically charged. Charged particles are unable to move across magnetic-field lines. Electromagnetic forces compel them to follow tight spiral paths along the field lines. Therefore, they primarily enter the atmosphere along the lines of the earth's magnetic field. When they collide with atmospheric molecules, they cause the latter to glow, resulting in polar aurorae.

In April 1958, the physicist James A. Van Allen of Iowa State University made an amazing discovery. Charged-particle counters were incorporated in the first of the *Explorer*-series satellites. They were measuring instruments similar to Geiger counters, which are used on earth to measure the strength of radioactivity. At a specific height above the earth, the counters first registered an extremely high number of particles, and then hardly any. The same phenomenon was found at the same time by counters in Soviet satellites. Van Allen had a brilliant idea that explained the behavior of the counters. Normally the electronics registered every particle captured by the Geiger tubes. Van Allen recognized, however, that the electronics would fail if *too many* particles arrived in too short a period of time. He concluded that the counters were not recording particles, not because no particles existed there, but because there were far too many of them. Van Allen was right. And since then, the earth's radiation belts are known as *Van Allen belts*. The radiation intensities that humans would encounter in the earth's radiation belts are simply enormous. Each square centimeter of a person's skin would be exposed to 100,000 particles per second.

It had been known earlier that charged particles can follow highly complicated paths in magnetic fields — much more complicated ones than those of bodies under the influence of gravity. There are, for example, regions of magnetic field lines from which, once captured, a particle cannot escape. The fields are magnetic traps. Two belts of trapped particles exist around the earth, called the *inner* and *outer* radiation belts (Figure 32). The particles follow spiral paths from north to south and then back again. They avoid regions where the magnetic fields are too strong and are therefore reflected in the polar regions, where the field is stronger than elsewhere.

What is the source of these particles that are bound by the earth's magnetic-field lines and which are bounced back and forth like tennis balls between the magnetic poles? Many of them originate in the flow of particles from the sun that surrounds the earth's magnetic field. As the solar wind encounters the earth, it bends the magnetic-field lines back toward the side opposite to the sun and draws them far out into space, forming the earth's long magnetic tail mentioned earlier. The eventful journey of the *ISEE-3* space probe described in Chapter 2 took it twice into the earth's magnetic tail. The solar wind also feeds charged particles, primarily electrons and protons, into our planet's radiation belts. But other particles often come from the earth itself. In the layers forming the ionosphere, ultraviolet light from the sun bombards atoms in the earth's upper atmosphere and removes negatively charged electrons. The remaining atoms are positively charged. They are known as *ions,* whence the name of the layers. Many of these particles travel out beyond the outermost layers of the atmosphere. There they are captured by the earth's magnetic field, which is being continually

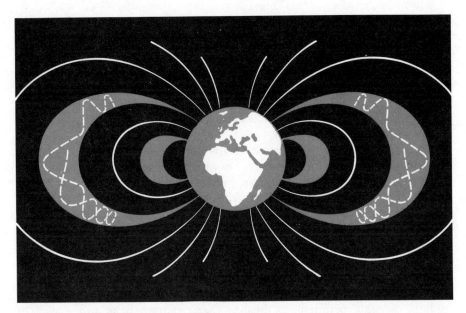

Figure 32. *The earth's radiation belts. Currents in the earth's interior produce the earth's magnetic field, some of whose field lines are indicated and which stretch far out into space. Electrically charged particles, such as electrons, are very mobile in magnetic fields. Their spiral paths along the magnetic field lines are shown here by the broken lines. The charged particles are, however, reflected by regions where the magnetic-field intensity increases, that is where the magnetic-field lines come close together. The electrons move in shells with a crescent-shaped cross section. Particles are reflected in regions near the poles. The polar regions act like mirrors, so that the particles constantly bounce backward and forward between the poles. Two radiation belts are known. The outer one lies at about 3.7 earth-radii in the equatorial plane, whereas the inner one is at about 1.6 earth-radii. Only the inner part of the earth's magnetic field is shown here. At greater distances from the earth, it is so weak that its shape is governed by the solar wind. The earth is therefore accompanied by a magnetic tail, which is turned away from the sun and stretches about one hundred earth-radii out into space. The ISEE-3 space probe (see Figure 20) passed through this region several times.*

compressed and distorted by the influence of the solar wind, and are so greatly accelerated that the electrons reach energies comparable to those used in medical x-ray equipment. With their high velocities, the charged particles originating in the ionosphere are able to reach the radiation belts, where they are trapped for some time.

Another component of the radiation in the Van Allen belts consists of particles, known as *cosmic rays,* that reach us from the depths of the Milky

Way. These are atoms with exceptionally high velocities, atomic nuclei that smash into the upper layers of the earth's atmosphere. Frequently, these collisions release neutrons, which are able to move freely through the magnetic field, because they are electrically neutral. But free neutrons have short lives. On the average, every free neutron decays after about 13 minutes into a positively charged proton and a negatively charged electron. These electrically charged particles are retained by the magnetic field. They are scarcely born before they are captured.

The particles are not necessarily retained in the belts forever. When the whole magnetic field is disturbed, many of the trapped particles escape and cascade into the earth's atmosphere, producing polar aurorae. Such disturbances may arise when dense clouds of gas in the solar wind encounter the earth's magnetic field.

The *Argus* experiment mentioned earlier injected electrons (and other charged particles) into the earth's magnetic field. They spread out along the field lines, and the majority of them were trapped in the belts for a long time. The earth was soon surrounded by a shell of electrons trapped by the magnetic field. Van Allen had shown that the earth has two radiation belts, and the *Argus* experiment established that radiation belts and a shell of radiation can be created artificially. Those were innocent times, when people played around in space with atom bombs. In 1963, the superpowers agreed to ban nuclear explosions in space.

We shall see in Chapter 9 that other planets, particularly Jupiter, have radiation belts. On earth we can detect radio emission from the charged particles trapped by Jupiter. The earth's radiation belts produce powerful radio emission as well, with wavelengths mainly in the region between 100 and 300 meters. Luckily, the ionosphere does not allow these wavelengths to reach the ground, otherwise the radio noise from the radiation belts would overwhelm all transmissions in the medium- and long-wave bands. Radio Earth can occasionally radiate 200,000 times as strongly as a typical medium-wave station, which may have an output of 500 kilowatts.

▪ *The Ring around the Earth*

Although no one is nowadays inclined to repeat explosions of atom bombs in space, pollution of our immediate environment in space is proceeding apace. At present, about 6000 artificial objects larger than 10 centimeters in diameter are orbiting the earth. In addition, there are some 40,000 pieces of roughly 1 centimeter in size and innumerable tinier pieces that are remnants of expended rockets and other material: veritable space garbage.

The zone where the so-called geostationary satellites orbit is particularly densely populated. These satellites follow a circular orbit at a height of about 5.6 earth radii above the equator. At this altitude, the satellites take 24 hours to orbit the earth. Thus, once injected in the correct direction, a satellite will remain over the same spot on the rotating earth. This is important for communications satellites, for example, which must hover over the Atlantic to provide telephone links between America and Europe. Similarly, broadcasting satellites are in geostationary orbits and therefore appear to remain at a fixed point in the sky when seen from earth. More than 300 satellites are in geostationary orbit at present, and more are added every year.

It has long been known that Saturn has rings, and more recently we have found that Jupiter and Uranus are surrounded by rings also (see p. 182 and p. 217). Neptune, too, has a ring (see p. 229). And now, in the last few decades, man has surrounded the earth with a ring of space debris. It may become even worse. In Florida, engineers and the proprietors of funeral parlors have set up the Celestis company, which wants to launch funeral urns into space for a $3900 fee. Using a special process, the ashes of the departed will be compressed into capsules of between 1- and 3-centimeter diameter. A total of 15,000 such capsules would then be packed into a container sent up into orbit. This space mausoleum will be fitted with a highly reflective surface, so that Junior with his homemade telescope can watch reverently as Grandfather rises brilliantly in the west.

▪ How Old is the Earth?

It is typical of the way in which people think that they always try to imagine how things began, and usually astronomy has the right answers. The earth is unbelievably old, but it was created a finite time ago. In Chapter 12, we shall examine the current view of the genesis of the planets. Right now, we shall only discuss *when* it happened.

Quite early on, Edmond Halley tried to estimate the age of the earth from the amount of salt in the sea. Rivers transport salt to the oceans, where the water evaporates, leaving the salt behind. The salt does not take part in the water's endless cycle, but makes a one-way trip to the sea. If one knows how much salt the rivers carry to the seas each year, then one can estimate over how long a period the sea has been getting more salty. The salt content seems to increase by about one millionth part of 1 percent each year. The current value of 3.5 percent would be reached after a few hundred million years, which is far too short in comparison with the age of the earth generally accepted today.

For several reasons, estimating the age of the earth from the salt-content of the sea is not a good method. It assumes, for instance, that the rivers have always transported the same amount of salt each year. In the past, however, mountains have been formed and have later vanished, and rivers have cut their way through this continually changing landscape. We do not know how much salt they carried previously. If we assume that in the past salt transport was about one one-hundredth of current values, then the age of the earth amounts to a few billion years.

A better method of estimating the earth's age is described in detail in Appendix B. Certain chemical elements undergo radioactive decay with time, many of them requiring millions or billions of years for a significant fraction to turn into other elements. The decay of these elements can be used as a clock to measure the time since the earth was formed. Terrestrial rock samples examined in this way show ages that are of the order of billions of years. We now believe that the earth was formed about 4.5 billion years ago. About 1 billion years later, primitive forms of life appeared.

▪ Danger from Space?

The idea of an astronomer discovering a celestial body out in space that turns out to be heading straight for the earth has been exploited by many good science-fiction writers — and by numerous bad ones as well. But bodies that arrive from space and create devastation by their impact are not just fantasies of imaginative authors. One such object fell in Siberia at the beginning of this century.

The peasant farmer I. P. Semenov was sitting on the steps of his house when he saw a brilliant burst of light in the north. He felt such a surge of heat that he was frightened his clothes would catch fire. Then the shock wave arrived. The floor of the house was ripped up, window panes shattered, and iron roof braces were broken. Semenov was thrown from the steps and lost consciousness. He was not the only witness, however. At 7:17 A.M. on 30 June 1908, the sky in the middle of Siberia was lit up as if by a second sun that was traveling from south to north. Then came a thunderclap that caused the engineer on the Trans-Siberian Railway to bring his train to a halt, assuming there had been an explosion close to, or aboard, the train. The seismometers at various institutes in Russia registered strong earth tremors, and the seismic waves from the Siberian forest were recorded even as far away as Jena in Germany. It was never established how many nomads were killed in that split second. Thirteen years later, the first Soviet researcher investigated the event, and an expedition reached the area only in 1927. The devastation was still clearly visible. Trees had been snapped off halfway up their trunks, others had been completely uprooted, and signs of a giant fire were found. A marsh that was nearly 10 kilometers across must have been the center of the impact, because the trees in a wide surrounding area were lying with their tops directed away from that spot. On that morning in 1908, a body some hundred meters in diameter must have hit the earth from space. The earth does not move in empty space, and from time to time something does fall onto its surface from outside. Since any body arriving from space must have a velocity of at least 11 kilometers per second, such an event can be catastrophic, unless the body is so small that it is strongly decelerated as it enters the atmosphere and at least partly vaporized.

In the past, there have been numerous such events, some of which have been even more powerful than the one in Siberia. About 15 million years ago, a body fell in Europe that produced a crater about 25 kilometers across. Today the town of Nördlingen, Germany, lies in its center. The crater has long since been filled up, but the original crater rim still forms a ring surrounding the town. Enormous amounts of energy must have been released by its impact. Ejected material was flung as far away as Bohemia, a distance of more than 150 kilometers. The surface rocks were melted by the impact and mixed with material from the impacting body, forming a speckled stone, called suevite. The tower of a church in Nördlingen was built from this material — to my knowledge the only church tower that arrived directly from heaven. A large meteor crater in Arizona bears witness that a massive body fell onto the earth some 20,000 to 25,000 years ago. Old impact craters that cannot be recognized from the Earth's surface are still being found from aircraft and satellites.

Fairly recently, the theory has been advanced that impacts from space occur in regular showers. The intervals between the episodes are supposed to be around 26 million years. This immediately inspired many scientists to speculate about the consequences of such recurrent catastrophes. Cosmic collisions have even been blamed for the puzzling extinction of the dinosaurs. But more about this in Chapter 8.

There is no lack of cosmic material that has fallen onto the earth. Some consists of meteorites, the stony and iron-bearing bodies that have fallen, and continue to fall daily, onto the earth. In addition, tons of dust arrive on the earth from space. In areas well away from industrial pollution, as in the Arctic, such dust can be found in rainwater and in snow.

It appears that the earth not only captures fine particles and larger bodies from the depths of space; surprisingly enough, some stones recovered from the Antarctic ice actually came from the moon.

FOUR

The Moon, Our Lifeless Neighbor

In this year . . . after sunset when the moon had first become visible a marvellous phenomenon was witnessed by some five or more men. . . . Now there was a bright new moon . . . and suddenly the upper horn split in two. From the midpoint of this division a flaming torch sprang up, spewing out, over a considerable distance, fire, hot coals, and sparks.

GERVASE OF CANTERBURY
The Chronicle of the Reigns of Stephen, Henry II., and Richard I.

When, in 1609, Galileo Galilei directed his telescope onto the moon, he realized that "the surface of the moon is not completely smooth, free from any unevenness and truly spherical, as one school of philosophers maintains, but is instead very irregular, covered in hollows and hills, just like the surface of the earth, on which high mountains and deep valleys can be found everywhere."

On 31 July 1964, the American probe *Ranger 7* crashed into the surface of the moon in the area since named Mare Cognitum. Before it smashed to pieces, it sent back more than 4000 close-up pictures of the moon, showing that it is saturated with craters, the smallest of which had diameters of one meter and less.

On 20 July 1969, an *Apollo* lander carried the first men to the barren cratered surface of the moon. The exploration of the earth's satellite had reached its peak.

▪ *The Earth's Satellite*

The moon is the sole celestial body that kept its role in the change from the Ptolemaic to the Copernican system: It remained the body that orbits the earth. Its path is an ellipse. This is only approximately correct, however, since celestial bodies move in elliptical orbits only in the gravitational field of a single, spherical body. The moon, however, is revolving around an earth that is flattened, rather than spherical, and the equatorial bulge also exerts a pull on the moon. The attraction between the earth and the moon is more complicated than it would be with a simple sphere. In addition, the moon does not just move in the gravitational field of *one* body, because the sun, as well as the earth, affects it. It is thus moving in the gravitational fields of two bodies. (If we want to be precise, we would say that it is also affected by the attraction exerted by the other planets. But since their gravitational forces on the moon are very small in comparison with those of the earth and the sun, we can ignore them.) But even with the alternating forces of just two bodies, its motion is extremely complicated. It was always a very difficult task for astronomers to calculate the moon's future position. More recently, with the aid of computers, the calculations have not actually become simpler but faster and a lot more convenient to execute.

Not only is it difficult to calculate the moon's motion, it is complicated even to describe it. We shall try to do so by envisaging it as consisting of a simple motion in an ellipse, combined with an additional, much slower motion (see Figure 33). Just as the orbit of the earth around the sun lies, to a good approximation, in a plane, so does the orbit of the moon around the earth. We can therefore talk about two orbital planes, that of the earth and that of the moon. These two planes are not identical, the moon's being inclined to the plane of the earth at an angle of about 5 degrees. The line of intersection of the two planes is known to astronomers as the *line of the nodes*.

If the earth were to follow just its elliptical orbit in its orbital plane around the sun, and the moon its elliptical orbit in its own orbital plane around the earth, all would be simple. But perturbations caused by the flattening of the earth and by the sun's gravitational attraction complicate the motions. Although the moon's orbit closely approximates to a plane, after orbiting the earth once it does not come back to precisely the same place it occupied previously. After a month, it is not in exactly the same plane. The motion can be described approximately if one imagines that as the moon revolves, its orbital plane slowly rotates in space. The line of the nodes, or the intersection of the earth's and the moon's orbital planes, moves. This gradual motion is in the direction opposite to the moon's orbital motion around the earth (Figure 33). It completes one rotation in the course of 18 years and 224 days. This is not the only complication in the moon's motion, but we shall not discuss the others here.

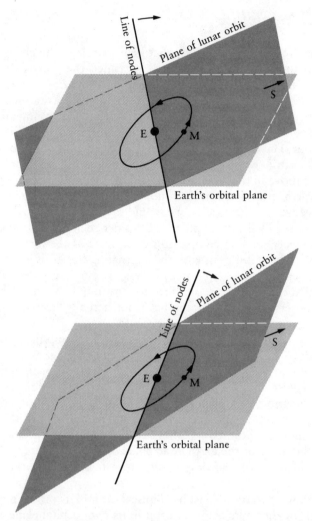

Figure 33. *The plane of the moon's orbit around the earth is inclined at about 5 degrees to that of the earth around the sun. The line where the two intersect is known as the* line of nodes. *The sun perturbs the motion of the moon, causing the line of nodes to rotate. It completes one rotation in about 19 years.*

▪ Full, Half, and New Moon

In the following discussion, we shall ignore the fact that the planes of the moon's and earth's orbits are inclined at a small angle to one another. So let us simplify things and assume that the earth and the moon move in the same plane, as shown in Figure 34.

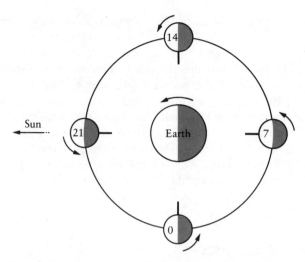

Figure 34. *From the earth, we see differing amounts of the half of the moon that is illuminated by sunlight, according to the position of the moon relative to the earth and the sun. Earth and moon are not shown to scale in this diagram. During the course of one revolution around the earth, the moon rotates so that a flagpole on the surface (indicated by a short line), pointing toward the earth at the beginning of a month, continues to do so. The time (in days) that has elapsed since the moon was at the bottom position is shown for each of the four intervals. It will be seen that for half an orbit, the flagpole will be on the day side of the moon.*

The three bodies—sun, earth, and moon—may have different relative positions according to the position of the moon. Figure 34 shows four of these positions and also explains how the different phases of the moon arise, depending on how we on the Earth see the illuminated half of the moon from different directions.

From the illustration, we see when the rounded side of the crescent moon is toward the left, new moon may be expected shortly; when the crescent is toward the right, new moon has passed, and the moon is waxing. This, of course, only applies in the northern hemisphere.

Let us look at the position of the moon at new moon again and imagine ourselves on its night side. We would then see the earth hanging in the dark sky on the opposite side of the heavens to the sun, and it would appear to be fully illuminated by the sun; we would see "full earth." Thus, full earth coincides with our new moon. Since the earth seen from the moon appears much larger than the moon seen from the earth, we would see a giant, fully sunlit disk. Night on the moon at the time of full earth is much brighter than full moon on earth. We can see this even from earth. Full earth coincides with our new moon and occurs during the daytime, when we are unable to see the moon. But when the small waxing crescent moon is visible in the evening twilight a few days later, or the waning crescent in the morning sky a few days earlier, it is not quite full earth on the moon, yet the illumination on the moon is still very bright. From the earth, we see not only the moon's small, fully sunlit crescent, but also the side in shadow, which appears faintly illuminated. The whole of the moon's disk can be seen, part brilliantly illuminated by the sun and part more faintly by the (nearly) full earth. This illumination is known as *earthshine*. The light was emitted by the sun, reflected by the earth onto the moon, and then reflected back again to the earth. Alexander von Humboldt (1769–1859) called it the "reflection of a reflection" in his famous book *Kosmos* and also mentions that it had been correctly explained by Leonardo da Vinci (1452–1519).

▪ The Canon of Eclipses

It must have been a source of great concern to ancient peoples when a shadow covered the bright disk of the full moon, causing it to become so dark that perhaps only a coppery glow remained. They recorded such events, with the oldest accounts of lunar eclipses dating back to the years 2283 and 1136 B.C. One record comes from the city of Ur in southern Babylonia, the other from China.

People must have been even more concerned when a bite was suddenly taken out the circular disk of the sun, which then shrank to a crescent and sometimes even disappeared entirely behind the moon's disk. It became dark in the middle of the day, and the landscape was only illuminated by the pale light from the rays of the corona surrounding the darkened disk of the sun.

If the orbit of the moon were exactly as shown in Figure 34, then at new moon the moon would always lie in front of the sun as seen from earth, and would partially or completely cover it; we would have a solar eclipse every month. Half a month later, at full moon, the earth would lie between the sun and the moon, the Earth's shadow would cross the moon, and we would have a lunar eclipse every month. However, eclipses do not occur every month because of the inclination of the moon's orbit to that of the earth, as shown in Figure 33. In general, the new moon is not directly in front of the sun, and the full moon is not precisely on the sun–earth line, so eclipses do not normally occur. Only when the moon is close to the line of the nodes and the latter simultaneously points toward the sun, can an eclipse occur. The moon then either appears close to the sun or on the opposite side of the sky from the sun. In the first case, we have new moon, in the second, full moon.

Let us initially assume that the plane of the moon's orbit is fixed in space. In the course of the year, eclipses can then occur on two occasions, separated by six months (Figure 35). This diagram shows that favorable conditions for an eclipse would occur at the same times each year. As we saw earlier, however, the line of the nodes rotates once in about 18.6 years. The two opportunities for eclipses therefore come about 20 days earlier each year. Whether an eclipse does occur during one of these two favorable periods depends on whether the moon passes close to the line of nodes at the appropriate time. This recurs after 6585.3 days, which is 18 years and 10½ or (11½) days in our reckoning. The difference of one day arises from whether there have been five or four leap years within the preceding interval (Figure 36). This is the Saros cycle of eclipses, which was known to the Babylonians (see Chapter 1).

Every 6585.3 days the moon is not only in its own and the earth's orbital planes (because it is on the line of the nodes, whose points lie on both planes), but it is also at the same phase, namely new or full moon. At new moon, it completely or partly hides the sun for a short while, when we observe a total or partial *solar* eclipse. At full moon, the earth's shadow falls onto the moon, and depending on how far the moon comes within the earth's shadow,[1] we observe a total or partial *lunar* eclipse.

Anyone who accurately calculates the movements of the earth and the moon is not only able to predict eclipses and say when eclipses occurred in the past but can also determine the places on the earth where an eclipse will appear total, where it will be only partial, and also when it will begin and end.

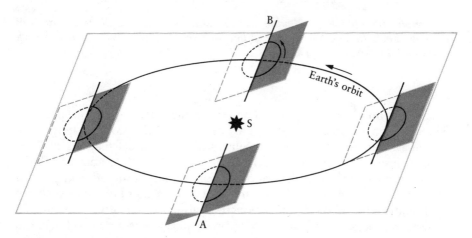

Figure 35. *The position of the plane of the moon's orbit at various times during the year. Eclipses can only occur when the earth is at positions A or B, because then the line of nodes points toward the sun. However, an eclipse can occur only when the moon is close to its line of nodes at such a time. It is then either full or new moon.*

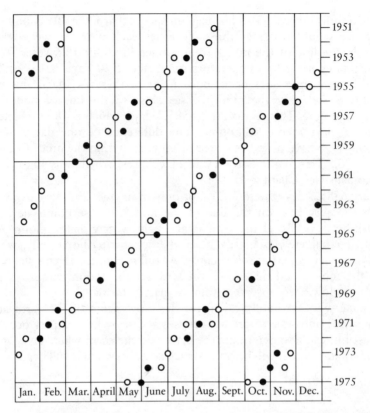

Figure 36. *Solar (circles) and lunar eclipses (dots) occur at two distinct periods in the year. These times come slightly earlier each year, and the cycle repeats after 18 years and 11 days.*

The man who played an outstanding part in the calculation of eclipses was Theodor von Oppolzer (1841–1886) of Vienna. In 1887, this physician-turned-astronomer published a volume with detailed calculations of 8000 solar eclipses and 5200 lunar eclipses between 1207 B.C. and A.D. 2161, including maps indicating the limits of their visibility from different parts of the earth. The result was a thick tome entitled *Canon of Eclipses,* which can be found in the libraries of all long-established observatories. Figure 37 shows a reproduction of Oppolzer's map of solar eclipses that will occur in the next few decades.

Columbus's eclipse described in the Introduction can be found in Oppolzer's *Canon* as number 4187. On the other hand, one may seek in vain for the eclipse described in Mark Twain's *A Connecticut Yankee in King Arthur's Court.* No eclipse took place on 21 June 528 B.C. as Mark Twain maintained—it was more than ten days to the next new moon.

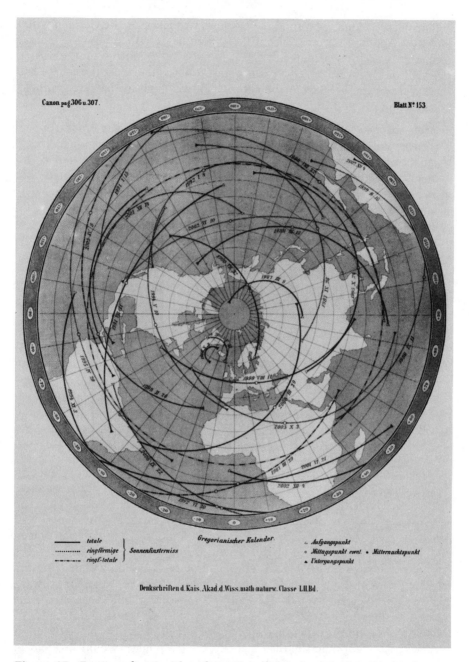

Figure 37. *Future eclipses, taken from Oppolzer's* Canon of Eclipses *of 1887. The circular arcs show the positions on the earth at which a solar eclipse may be observed on the appropriate date. The solid lines indicate total eclipses. The* Canon *shows that a total eclipse will cross Hawaii, parts of Mexico, and the northern part of South America in July 1991.*

The finest account of a total solar eclipse was given by the German writer and poet Adalbert Stifter (1805–1868), when he described how the moon's umbra crossed Vienna on 8 July 1842. He not only reported on the natural phenomena that were observed, but also recounted the effect the eclipse had on people as the light from the sun, which was high in the sky, gradually became dimmer and dimmer, before finally vanishing:

> After the first, shocked silence, there arose an inarticulate sound of wonder and amazement: One held up his hands, another wrung them in quiet agitation, while yet another clutched them together fiercely—one woman began to cry violently, another in the house next door fainted; and a man, a very grave, strong man, later told me that tears were running down his face. . . . Never in my life have I been so moved, so moved by any such spectacle and such sublimity, as in those two minutes—it was just as if God had suddenly spoken, and I had understood. I descended from my viewpoint just as Moses descended, thousands and thousands of years ago, from the fiery mountain; confused and dazed.

I myself have been lucky enough to see the total solar eclipse that was visible in Italy and Yugoslavia on 15 February 1961. As an astronomer, I expected, of course, to observe such an event quite dispassionately—after all, there is nothing mysterious about the shadow of the moon. But I have to admit that I was no different from that "grave, strong man."

▪ The Unchanging Face of the Moon

As the moon revolves once around the earth in the course of a month, it also rotates once around its axis, so that it always turns the same side toward us. This is no accident. Let us assume that the moon's rotation was at one time much faster. In that case, the tides raised by the earth would have slowed down its rotation, particularly while it was still molten. Its rotational energy would have been dissipated by tidal friction, and it would have rotated more and more slowly around its axis, until it came to rotate only once for every revolution around the earth. It would then always turn the same face toward the earth. To an observer on the moon, the two tidal bulges would always remain in the same places, one exactly where the earth was in the zenith of the lunar sky, and the other on the opposite side of the moon. Since the earth would remain motionless in the lunar sky, the tidal bulges would also remain stationary; the friction would have come to an end and would no longer brake the rotation of the moon.

This condition was reached a very long time ago in the moon's history. There were no oceans with tides to slow down the moon. The whole body of the moon was deformed by the tides. This continuous kneading of its material was braking its rotation. For this reason, we see year in and year out the same half of the moon and exactly the same "man in the moon."

Because of the moon's strange rotation, a lunar day lasts 29 earth-days. Although with respect to the fixed stars it rotates once around the earth in 27 days, thus keeping the same side toward the earth, it and the earth also move partly around the sun. Thus, a lunar day is longer than the time it takes to rotate around the earth. For more than 14 days, therefore, the surface dust on the moon is exposed to the full glare of the light from the sun, with no atmosphere to soften it, and no clouds to alleviate it. As a result, the dust is heated to about 120°C, before dropping to −190°C during the equally long lunar night.

Actually, this is not quite correct, because we do see slightly more than half of the moon's surface. Its rotation is perfectly uniform, but the speed at which it orbits around the earth is not constant. It moves around our planet in an ellipse. Therefore, although the ellipse is also very close to a circle, the moon is sometimes nearer to us, sometimes farther away. In accordance with Kepler's second law, the moon moves faster in its orbit when it is close to the earth, and slower when farther away, so its orbital velocity is not constant. Constant rotation and inconstant orbital velocity cause it to show more of its western side at times, and more of its eastern at others. Over the course of a month, it oscillates slightly back and forth as seen from the earth. Astronomers call this oscillation lunar *libration*. There are additional reasons why we see more than half of the moon's surface, but I shall not discuss them here.

▪ *An Airless World*

When it was realized that there were mountains on the moon, it was first thought that the apparently relatively flat, dark areas of the surface were seas. In 1686, the 26-year-old Bernard Le Bovier de Fontenelle (1657–1757) explained the nature of the universe to the Marquise G** in his *Entretiens sur la Pluralité des Mondes* (*Dialogues on the Plurality of Worlds*). In the German edition, which appeared nearly 100 years later, he said of the moon, "The famous Cassini,[2] who is a man who has made the most profound exploration of the heavens, has discovered something on the moon, that divides into two, then later reunites, and finally disappears into a form of shaft. We have every reason to suspect that this is a river." The Berlin astronomer Johann Elert Bode (1747–1826) commented on this in a footnote, "Hevel,[3] who of all astronomers most industriously observes the moon, and who regards the dark areas as definitely being seas, has, however, discovered nothing at all similar to a river."

Given that the moon was a body with seas, and perhaps even with rivers, the question arose whether forms of life could exist there. Georg Christoph

Lichtenberg, the Göttingen physicist and man of letters, said mockingly, "An astronomer knows whether the moon is inhabited with approximately the same probability as he knows who his father was, but not with the same degree of certainty as he knows who his mother was."

In a popular astronomy book of 1844, I found the assertion that one might not recognize any possible lunar inhabitants with telescopes:

> . . . therefore, a great astronomer has suggested another way of obtaining convincing evidence of the moon being inhabited by rational beings. As such, they certainly study geometry . . . and they must therefore necessarily be fully familiar with many of our theorems. If, at the time of a lunar eclipse — when the inhabitants of the side of the moon that is turned toward us have a relatively long solar eclipse and are looking at the night side of the earth — we were to depict a geometrical figure, such as Pythagoras' theorem, with long lines of fire on the night side . . . we would have thrown a gauntlet to the inhabitants of the moon that could only be picked up by equals. The selenites could best send back their reply at the time of a solar eclipse, and how utterly embarrassing it would be if they told us the solution to a problem that we had previously sought in vain!

But there is no life on the moon, because there is no atmosphere that life could breathe. This can be determined very simply. When the moon passes in front of a fixed star in the course of its passage across the sky, the star vanishes behind the moon's eastern side, without showing the slightest change in brightness beforehand. If the Moon were surrounded by a layer of air, then the light from the star, on its way to us, would pass through the gases in the lunar atmosphere shortly beforehand. It would be refracted, and the star would show a slight shift in position before it passed behind the edge of the moon. But this has never been observed.

In fact, the moon is unable to retain an atmosphere. In the case of the earth, the escape velocity is 11.2 kilometers per second, but the moon's gravity cannot even retain bodies that move away from it faster than 2.4 kilometers per second. Even at room temperature, many molecules in the air have higher velocities and would therefore escape. If the moon ever had an atmosphere, it would certainly have lost all its surrounding shell of air in the subsequent billions of years of its lifetime.

▪ The Mountains of the Moon

Galileo may not have been the first to observe the moon through a telescope, but he was the first to see that the irregularities of light and shadow at half moon were mountains and craters.[4] He knew that the contrasts of light and shadow threw surface features into relief; and from the length of the shadows, he estimated the height of the mountains as 7000 meters. Nowadays, we are able to recognize from earth individual features on the moon down to a few hundred meters in size. Figure 38 shows a modern photograph of the crater Calvius taken from earth. We now know that the highest lunar mountains tower some 11 kilometers above the surrounding plains. The highest lunar mountains exceed the height of any mountains on earth.

This is not surprising. There is a rule of thumb for determining the height of mountains on any planetary body. It is governed by the fact that rocks cannot withstand unlimited pressure. At 1000 atmospheres, the hardest granite begins to flow. Since the whole universe consists of the same types of elements, we may expect the planets and their satellites to be formed of similar rocks. This has actually been confirmed by lunar samples. The unmanned landing craft on Mars and Venus also found rocks that are related to those on earth.

How high can a mountain be on any planet? The more rock we might try to pile up into a mountain on earth, the higher the pressure on its foundations becomes. If the mountain becomes too big, the layers beneath it become weak and begin to yield. With every added stone, our mountain would sink deeper into the ground. The higher a planetary body's surface gravity, the greater the pressure a mountain exerts on the underlying material, and the sooner the ground will give way. On earth, the highest mountains

Figure 38. *This photograph taken from earth shows Clavius, a lunar crater more than 200 kilometers across. Smaller craters have been formed in its interior at a later stage* *(Photo: Palomar Observatory).*

cannot be higher than 10 kilometers—even Mount Everest, at 8848 meters, complies with this rule. On the moon, gravity is only one-sixth of that on earth. Thus, a lunar mountain can be six times the height, 60 kilometers, before the lunar surface gives way. So we should not be so surprised that lunar mountains exceed the height of those on earth—we should perhaps ask why mountains on the moon are not even higher. This is doubtless related to the way in which the lunar mountains were formed.

▪ Luna, Lunar Orbiter, Surveyor, and Apollo

Although the telescope showed many features on the moon—the craters, the mountain ridges, the dark plains called "seas" and usually known by the Latin "maria," and the long narrow valleys ("rilles") in the surface— conclusive details were only obtained when artificial robots and eventually man reached the moon.

The development of this particular type of lunar research lasted 17 years. It began on 7 October 1959 between 6:30 A.M. and 7:10 A.M. Moscow time, when the Soviet probe *Luna 3* photographed the far side of the moon as it flew past it and later transmitted the pictures back to earth over distances of as much as 470,000 kilometers. The United States only caught up in 1966, when the *Lunar-Orbiter* program with five probes in lunar orbit enabled the compilation of a photographic map that covered 95 percent of the lunar surface and showed features with diameters as small as about 1 meter.

I still remember the general assembly of the International Astronomical Union in Prague in 1967. American astronomers had covered the floor of an otherwise empty room in Prague University with a mosaic of photographs, which together formed a most impressive picture of the entire lunar surface. Astronomers who had traveled to Prague from all over the world were creeping around in stocking feet over this cratered landscape, which was otherwise protected only by a thin sheet of plastic.

In February 1966, the Soviet probe *Luna 9* landed safely on the moon and transmitted the first photographs of its landing site. By January 1968, a further *Luna* probe and five American *Surveyor* probes had soft-landed on the lunar surface, had reported about the nature of the soil, and had analyzed the rocks. In 1968 and 1969, American astronauts orbited the moon in preparation for a manned landing. In July 1969, the *Apollo-11* lander touched down on the Mare Tranquillitatis, the "Sea of Tranquillity." It returned to earth with 28 kilograms of lunar rocks. Between then and 1972, the six manned lunar expeditions of the *Apollo* program visited various types of surface features. Between 1970 and 1976, six Soviet robots in the *Luna* series made soft-landings on the moon, conducted tests on the surface, obtained core samples of the soil, and returned them safely to earth.

Since then, it has again been quiet on the moon. The results of the lunar missions have given us a lot of new information about the earth's satellite, but so far, they have not settled the controversy about the origin of the moon.

NASA's *Apollo* program ended in 1972 after six manned landings on the moon. Will the footprints left in the lunar dust by the *Apollo-17* astronauts, Eugene Cernan and Harrison Schmitt, be the last traces of man's activity for thousands of years? At present, no definite lunar missions are being considered in the United States, but at conferences the question of a permanently manned moon base is being discussed. Because of its lower gravity, it would be easier to supply minerals from the moon to an earth-orbiting space station. Plans for a 25-meter-diameter optical telescope have been discussed, as well as the question of obtaining oxygen from lunar rocks. The optimists among the planners hope that their ideas will bear fruit in the first decade of the twenty-first century.

■ *Volcanoes or Impact Craters?*

When Galileo recognized and drew the first lunar crater, he assumed that he was looking at an extinct volcano. The better telescopes became, the more craters could be seen on the moon. Robert Hooke (1635–1703), after whom the law of elasticity in physics is named, and a contemporary and adversary of Newton in numerous quarrels, summarized the rival explanations of the lunar craters in 1665. They either arose from the impacts of cosmic bodies or were lunar volcanoes. He wrote: "It would be difficult to imagine whence those bodies should come"—meaning the bodies that according to the impact theory must have collided with the moon. For this reason, he favored the theory of lunar volcanoes.

After Hooke, scholars did not spend much time pondering the origin of the lunar formations. For hundreds of years, the question was ignored. This is not surprising, because geology had yet to become a science. Any details of the earth's history were thought to be contained in the biblical account of the Flood, or else in the story of creation as given in the book of Genesis in the Old Testament. The only thing known about the moon was that God created it on the fourth day.

Until 1766, it was thought that mountains were destroyed in volcanic eruptions. In that year, Vesuvius erupted again, and people witnessed with their own eyes how it grew in the process. Obviously, volcanic eruptions could produce mountains. Were the lunar craters therefore actually extinct volcanoes as Hooke had suggested? The Phlegrean Fields (as the hilly, volcanic region near Naples is called) do actually resemble part of the lunar landscape.

Franz Aepinus (1724–1802) of Rostock, a councillor of state at the court of Catherine the Great, was one of the chief proponents of the volcanic theory. He pointed out how the central cones in many lunar craters greatly resembled the central cones in Vesuvius.

Today, after the *Apollo* missions have explored the moon, we know that almost all the lunar craters were caused by impacts. Even in recent times, stony bodies have crashed into the moon, their impacts melting and vaporizing both the bodies and the materials they encountered. The molten material was ejected over the lunar surface, quite frequently in preferential directions, causing the bright rays radiating from some craters. Billions of years ago, the impacts on the moon were much greater than in more recent times. Occasionally the impact craters became filled with originally fluid material from the lunar interior. The dark "seas" consist of lavas that were erupted from the moon's interior and filled whole basins, flooding craters with their dark material. Today, there are few craters in the dark maria, and we can see only those whose walls were high enough for the peaks to remain above the level of the lava flows, and those craters that have arisen from later impacts.

Impacts have produced not only many large craters but also innumerable smaller ones, down to sizes of less than one millimeter in diameter. In large impacts, the surface melted, but in smaller ones the rocks were broken into fragments, covering the moon's surface with fine dust. This dust layer is now at least one meter thick, but often its thickness reaches 20 meters.

The *Apollo* astronauts left their footprints behind in the lunar dust. But the dust also holds traces of events that occurred thousands, or perhaps even millions of years earlier. At many points on the moon, it is obvious that at some time or other, a block of rock rolled down a mountainside. It was probably dislodged by the shock of a "moonquake." The blocks can be seen lying in the valleys, together with the traces that they left in the lunar dust (Figure 39). Although much has been published about the results of lunar research, we have yet to see the headline "Rolling Stones on the Moon" appear in the tabloid press.

▪ The Moon in a Test Tube

The six *Apollo* missions brought back 382 kilograms of lunar rocks to our laboratories, while the three Soviet *Luna* robots returned with 310 grams of soil and cores. Although previously only terrestrial rocks and meteorites that had come from space could be chemically investigated in laboratories, lunar material is now available. The rocks from the moon turned out to be not so very different from terrestrial ones. The same chemical elements in similar ratios form similar compounds on the moon and produce similar crystals.

Figure 39. *This photograph of a region near the lunar crater Vitello, taken by* Lunar Orbiter *on 17 August 1967 from an altitude of 167 kilometers, shows (indicated by the arrows) where two rocks have rolled down a slope. On the left, the rock can be seen at the end of a wide track in the lunar dust. A narrower, longer track runs down the middle of the picture from the top toward the bottom. On the image, features as small as two meters across can be recognized (Photo: NASA).*

But there are differences. The moon has no water, not even bound up in rocks as on earth. In fact, it has never had any water. The samples returned from the moon show no signs of the existence of any organic life-forms, either now or previously.

It is not just the lack of water that distinguishes lunar rocks. Other chemical anomalies occur as well. The light, volatile elements such as sodium, calcium, potassium, chlorine, bromine, and mercury are rarer than on earth, whereas titanium, uranium, and thorium are more common.

By using uranium and thorium, one can determine the age of a rock (see Appendix B). The oldest lunar rock appears to have solidified about 4.6 billion years ago. The dark materials filling the mare impact basins cooled some three to four billion years ago. In one sample, the rock appeared to have solidified only about 600 million years ago. Thus, it would seem that there are craters that are less than one billion years in age.

▪ Watchmen on the Moon

The astronauts have left their lunar landing sites and have come back home. The Soviet robots are also no longer there. Luckily, the lunar missions did not just leave national flags behind but also useful scientific instruments that monitored the moon beyond the end of the landing missions. The seismometers left behind by the *Apollo* astronauts recorded faint moonquakes for eight years, transmitting their measurements back to earth. The moon has been set vibrating by the detonation of explosive charges, and by spent landing stages and rockets crashing into its surface. The instruments have also detected the vibrations caused when meteorites have fallen on its surface. But the equipment has also recorded true moonquakes, which give information about movements within its interior. The moon does not appear to be very active nowadays, and there are far fewer, and weaker, shocks than occur on earth. Moonquakes are somewhat more frequent when the moon is at the closest point of its orbit to the earth. It therefore appears that the earth's tidal forces are still at work, deforming the moon.

But the earth is not just responsible for causing moonquakes, it has also distorted the moon. Although the full moon may appear round, it is not a true sphere. Its shape is more like that of an egg, with the "sharp" end pointing towards the earth. Earlier, when the moon's interior was still fluid, the solid crust on the side turned towards the earth must have been thinner than that on the far side. The impact craters on the pointed side were mainly filled with lava that welled out of the impact sites, whereas on the far side, there are hardly any mare areas and therefore few lava flows.

We do not know whether the moon's core is still fluid. It has probably solidified right to the center. That might explain why it has no magnetic field. A compass would not show a lost astronaut on the moon the way home; but the stars could be used for that purpose, since they would neither be hidden by clouds nor be lost in the glare of a daytime sky.

The objects left behind by the *Apollo* astronauts included several reflectors. When a laser is beamed at these and the reflected signal is captured, the earth–moon distance can be determined from the transit time to within about 20 centimeters. The measurements provide a wealth of other useful material. The expected change in the orbit of the moon caused by tidal friction in the seas now actually appears to be confirmed by laser echoes from the moon.

▪ Where Did the Moon Come From?

Before man went to the moon, a well-known planetary researcher said, "Bring me a rock from the moon and I will tell you how the solar system was formed." He was brought a whole sackful of rocks from the moon, but despite this, we still do not know where the moon originated. For a long time, three possible processes have been considered that could have led to the earth–moon system in its present form. It had been hoped that landings on the moon would enable us to choose the correct explanation from these three possibilities. In this respect, the moon landings were a disappointment.

The three competing theories for the formation of the moon may be summarized as follows:

(1) The moon was born from the earth;

(2) the moon has been captured by the earth;

(3) the moon and the earth formed alongside one another.

The first theory was advanced by George Darwin (1845–1912), the son of the great Charles Darwin, founder of the theory of evolution. The younger Darwin envisaged the formation of the moon as follows: At least one billion years ago, there was just the earth. It rotated very fast, with the length of the day being only a few of our 24 hours. The centrifugal force at the equator was correspondingly high, and the earth was therefore far more flattened than it is now. Because of the great centrifugal force, the equatorial bulge developed a pronounced bump at one point. This bump grew larger and larger, until it finally separated from the earth. The future moon had been born. The body initially circled low over the equator, but tidal friction pushed it farther and farther out into space and away from the earth. At the same time, the tidal friction slowed down the earth's rotation. This continued until the current situation was reached in which the earth rotates once around its axis in 24 hours, and the moon itself orbits the earth once in 28 days.

Although the concept of the moon being torn from the entrails of the earth appears quite satisfying, it has several flaws. By using considerably simplified models, the splitting of the moon from the earth can be followed on a computer. In no case do just two bodies remain at the end of the process, as we find in the earth–moon system today. In the model calculations, the motions following the formation of the moon are quite different from those observed today. This is a point *against* the fracture theory. *For* it is the fact that the density of the moon is only 60 percent of the earth's mean density. The moon consists of lighter material than the average throughout the earth. But the earth's outer layers have low densities similar to those of the material forming the moon. Is the moon therefore part of the earth's outer layers? When the first *Apollo* astronauts brought lunar rocks back to earth, the whole situation became even more confused. The returned material, although very similar to terrestrial rocks, is deficient in light volatile elements. The abundance ratios between certain chemical elements are similar to those of terrestrial materials, but others are different. Tests on the lunar material did not contradict the fracture theory, but they did not confirm it, either.

Let us turn to the second possibility. The moon was originally a planet in the solar system, until the earth came too close, and the combination of the effects of the gravitational forces of the earth and the sun forced the moon into an orbit around the earth. (We have already seen in Chapter 2 that *one* body, such as the earth, is not sufficient to capture another body without the help of the sun.) The moon was, so to speak, adopted by the earth. In this case, the lunar material was derived from a completely different region of the solar system and should not resemble terrestrial material at all. But the moon does consist of material that is very similar to that found on the earth. This is confirmed, for example, by the abundance of various oxygen isotopes in the rocks. Oxygen atoms oxygen 16, oxygen 17, and oxygen 18 occur in lunar material with the same ratios as on Earth.[5] In meteorites, which do originate in other regions of the solar system (see Chapter 8), the oxygen ratios are completely different. Thus, we can say that lunar and terrestrial rocks are so chemically similar that they must have undergone a similar history. But they are not one and the same material.

There remains the third possibility. When the earth formed, the moon was born in its vicinity from similar material, that is, the earth and the moon are siblings. The material originally found in the earth's neighborhood must not have differed much from that from which the earth itself was formed. The question of the formation of the moon is related to the way in which the planets were formed. We shall consider this in Chapter 12.

▪ *Mysterious Lights*

In 1778, the newspaper *Neue Zürcher Zeitung* described a remarkable phenomenon that was observed by the commandant of the Spanish fleet, Admiral Don Antonio de Ulloa, from near Cape St. Vincent at the time of a total solar eclipse on 24 June 1778.[6] The total phase was visible in a narrow band that began over the Pacific. The path crossed Mexico, the southern part of North America, and the Atlantic, ending in North Africa.

The report says, ". . . before the sun could be seen again, a point of light was seen on the moon. . . . Don Antonio maintained that this point of light was a hole in the moon, through which the sun could be seen." According to the report, others also noted the same phenomenon.

In May 1783, William Herschel (1738–1822), the finest astronomical observer of his day—we shall encounter him again in Chapter 11 as the discoverer of the planet Uranus and say more about him there—saw a red point of light on the unilluminated portion of the moon. Herschel reported three further observations of glows on the dark side of the moon in 1787.

People were not very surprised by this, because they thought that the lunar craters were extinct volcanoes. Why should there not be a few active volcanoes as well?

The fact that occasional lava flows on the night side of the moon were possibly being seen from earth seemed to confirm the theory of the volcanic origin of the lunar crates. Reports are still received of reddish glows being seen momentarily on the dark side of the moon, as well as of occasional cloudy and misty patches at specific locations on the day side. If these effects are real, we do not understand them. Apart from them, we have seen no signs of volcanic activity. The moon appears to be solid to the core. How then could lava be erupted onto the surface?

Doubtless, not all the craters arose from impacts. There are, for example, lines of craters just like pearls on a string (Figure 40). Since impacting bodies do not arrive from space in Indian file, we must assume that these craters have been formed along a rift in the lunar crust. But although there has undoubtedly been volcanism on the moon, it has been extinct for a very long time.

What may have caused the various luminous phenomena seen on the moon? A possible explanation for the event described at the beginning of this chapter was given quite recently, but only after we became familiar with the far side of the moon.

Behind the eastern edge of the moon—invisible from earth—there is the crater Giordano Bruno, named after the scholar burned as a heretic at Rome in 1600. (Bruno went far beyond the beliefs of Copernicus in that he considered the fixed stars to be suns, perhaps even circled by inhabited planets; see Chapter 1.) Jack B. Hartung of New York State University at Stony Brook believes that a meteorite hit the moon on 18 June 1178, producing the crater Giordano Bruno, and that this event is the basis of Gervase of Canterbury's account (see the quotation that opens this chapter). When the body hit the

Figure 40. *These aligned lunar craters suggest that they were not caused by impacts like most of the others, but that they are of volcanic origin (Photo: NASA).*

moon, it not only excavated a crater in the surface, but the heat produced caused it and the impacted rocks to melt. Rocky fragments were partially melted and ejected high above the surface in a fiery shower. Although the site of the impact was on the invisible far side of the moon, the glowing cloud rose so high that it could be seen from earth above the edge of the moon. Massive clouds of dust must have been created, spreading far over the airless lunar landscape before sinking to the ground.

Hartung's interpretation of Gervase of Canterbury's account does have other possible confirmation. If rocks are ejected into space during such an event and do not fall back onto the moon, some of them may encounter the earth. Rock fragments can also be injected into an orbit around the sun that is very similar to the earth's. People have tried to calculate the approximate paths and found that fragments that immediately encounter the earth's gravitational field will be forced to enter the atmosphere in the daytime. Even though they may burn up as meteors (see Chapter 8), they will obviously not be seen at that time of day. Those that are injected into orbits around the sun, however, move in such a way that they will be captured by the earth a few months later. Thus, the earth must have picked up some of them. And the fragments falling on the earth must have produced a shower of meteors as they encountered the earth's atmosphere. In fact, an unusually rich meteor shower was observed in China, Korea, and Japan on 11 October of the same year. As a Korean chronicle put it: "Numerous stars flew towards the west." Perhaps these consisted of material from the moon that was ejected into space when the crater Giordano Bruno was formed, some fragments of which were later captured by the earth.

▪ Three Stones from the Moon

Long before man had set foot on the moon, an idea had been circulating: If a body crashes onto the moon, it will not just punch a hole in the lunar surface. Material will also be hurled upward. As we saw above, the moon's escape velocity is 2.4 kilometers per second. Therefore, when a body impacts on the moon and hurls material out with a greater velocity, it does not fall back but instead continues out into space. Some of the blocks of material escaping from the moon's gravity will be encountered by the earth. It is estimated that the earth captures between 10 and 100 tons of lunar material every year. Where are such stones to be found? Since they must have arrived as meteorites, many of them ought to be found in meteorite collections. But how can they be recognized and distinguished from other meteorites?

Only when astronauts brought lunar rocks back to earth could anyone hope to be able to use chemical analysis to differentiate between material from the moon and meteorites that came from other regions of the solar system. Success followed soon after.

In 1982, one of NASA's geologists examined a 31-gram meteorite that had been collected, with other meteorites, in the Antarctic in January 1982. Repeated chemical analysis showed that its composition was significantly different from that of the other meteorites. For example, it contained only about a third of the manganese in them, which is typical of samples from the moon and the earth.

People then began to comb through every meteorite collection for lunar rocks. They struck pay dirt in March 1984. A meteorite that had been found in the Yamato Mountains in Antarctica in 1979 closely resembled the lunar rock discovered in 1982. And yet another meteorite from the same location showed all the characteristics of lunar material.

We already mentioned the characteristic lunar and terrestrial oxygen ratios. The three rocks that probably originated on the moon have the same oxygen ratios as the moon and the earth. Since they arrived from space—as can be seen from the molten crust formed as they entered the atmosphere—they must have come from the moon.

FIVE

The Hot Planets

> The only planets on which we can expect with some
> degree of certainty that there is organised life, are
> Mars and Venus. Looking to the future, only Venus
> is moving toward an era when life will flower, be-
> cause there, as the flow of warmth from the sun
> declines with the passing of time, suitable conditions
> for higher animals will persist for a long period.
>
> SVANTE ARRHENIUSS (1859–1927)

■

Closest to the sun, which rules over the whole planetary system, is the orbit
of the planet Mercury. Since the time of Copernicus, it has been recognized
as the innermost planet. On 10 February 1860, however, the *New York
Times* congratulated two astronomers, one an amateur and the other a
professional of high standing, on their great discovery of a planet that orbited
the sun within the orbit of Mercury.

■ *The Hunt for Vulcan*

The great French astronomer, Urbain Jean Joseph Le Verrier (1811–1877)
had predicted the existence of the new planet in 1859. When Dr. Lescar-
bault, a physician and amateur astronomer in France, heard of this, he

remembered his own observation of the sun on 26 March 1859, when with his telescope, he had seen a small, round, black body cross the solar disk. But Benjamin Scott, Chamberlain of London, disputed his claim to fame over the discovery. He announced that he had already seen a body in front of the solar disk in the summer of 1847. He was even able to produce two witnesses, although one of them was his son, who was only five years old at the time of the observation.

The reason for the suspicion that there might be another inner planet was again the planet that caused many problems for those dealing with celestial mechanics: Mercury. Its orbit was an ellipse, as had been expected for a planet since Kepler's time, but somehow it did not seem to obey the simple laws. As it orbited the sun, it occasionally crossed in front of the solar disk. Such transits of Mercury, unlike those of Venus, could not be used to determine the earth–sun distance (see the discussion in Chapter 3), but they did offer a good opportunity to check Mercury's motion. Transits of Mercury predicted by orbital calculations occasionally occurred as much as a day, or at the very least several hours, late. When Le Verrier again calculated the shape and position of Mercury's elliptical orbit, predictions of its motion improved, but differences remained. Mercury was not behaving as Newton's law of gravity demanded.

One can get an approximate idea of its complicated orbital motion by imagining that as it traces an 88-day ellipse around the sun, its elliptical orbit slowly rotates in space. This occurs at such a slow rate that it takes 227,000 years for it to return to its original position (Figure 41). The other planets perturb Mercury and cause its orbit to rotate; however, it should take 244,000 years to complete one rotation. Where did the difference between 227,000 and 244,000 years come from? Le Verrier knew a way out. Only recently, he had dramatically solved a similar problem. We shall meet Le Verrier again in Chapter 11 as one of the two men who predicted the existence of a previously unknown planet: Neptune. Irregularities in the motion of Uranus had indicated perturbations caused by another body, which turned out to be Neptune. It is difficult for scientists to avoid the temptation of repeating tactics that brought them earlier success. Might not the irregularities in the motion of Mercury also be explained by a yet unknown planet?

Following his success in explaining the perturbations of Uranus, Le Verrier tried to explain the rotation of Mercury's elliptical orbit as being caused by the effects of a planet orbiting closer to the sun than Mercury. He named it "Vulcan." That nobody had yet detected it was not surprising. As seen from the earth, Vulcan would never be more than "a stone's throw" from the sun. It would therefore be in the sky in daylight and would rise and set immediately before or after the sun. After Le Verrier drew attention to Vulcan, a lot of people thought they had already seen it, mostly as it transited the sun. Others claimed that they had noticed it during twilight. In the end, not one of these observations was confirmed.

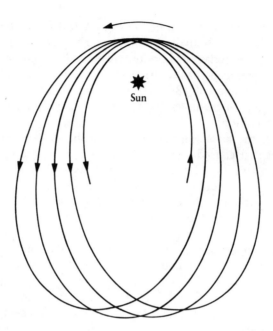

Figure 41. *Although Newtonian mechanics constrains a planet to move in an elliptical orbit around the sun, the orbit of Mercury is more complicated. It does move approximately in an ellipse with the sun at one focus, but this ellipse appears to rotate in space. Although this rotation is so slow that the planet's orbit requires about 227,000 years to complete a single turn, the motion has been known for a long time. The diagram is not to scale. In reality, Mercury's orbit is much closer to a circle, and the rotation of the orbit is far slower than indicated here.*

In 1876, a year before Le Verrier's death, a round object was seen to pass in front of the sun. Le Verrier used this observation to predict a further transit. Amateur and professional astronomers all over the world turned their telescopes on the sun, and waited to see if a small, round, black body would cross the disk. But Vulcan failed to turn up. Until he died in 1877, Le Verrier believed in a planet within the orbit of Mercury. Reports frequently appeared in newspapers prematurely celebrating the discovery of Vulcan. But time and again, the skeptics had their say. Finally, the critics won: there was no such planet as Vulcan. What, however, caused the rotation of the orbit of Mercury? Astronomers had to wait until 1915 for the solution to the riddle.

It required a scientist of stature comparable to Isaac Newton, namely Albert Einstein, to solve the problem of the rotation of Mercury's orbit. In his general theory of relativity, he showed that Newton's law of gravity only applies when gravitational fields are not particularly strong. Close to the sun, however, gravity is much stronger, and departures from Newton's law of gravity become noticeable. Mercury does move in an elliptical orbit, as it must according to Newton, but the perturbations caused by the other planets, which make that ellipse rotate around the sun, cannot account completely for the observed rate of the rotation. Thus, Mercury's apparent waywardness is based on laws discovered by Einstein that do not follow Newtonian theory.

Herr Meyer on Gallia

After Herr Meyer had read Jules Verne's story "Off on a Comet" in just a few days, he could not stop thinking about the people whom fate had marooned on Comet "Gallia" and who were forced to journey with it through the planetary system. In many of his dreams, Herr Meyer found himself involved with this involuntary journey. Since he had also just been reading details of recent investigations of the planets by space probes, the story by an author who had been dead for 80 years became interwoven with modern astronomical knowledge.

Herr Meyer found himself in a faintly lit room with a large glass window. Outside, a deep black, rolling landscape lay under a milky-white sky in which the sun stood high above the horizon. But what a sun! It was twice the size it appears to us. Unmoving, pinkish tongues of flame could be seen around its edge, silhouetted against the sky.

Herr Meyer realized that he was on Comet Gallia. Although he had not the slightest idea of how he had come to be there, he realized that he was on a tiny celestial body, flying through space. Thus, he was not surprised when the binoculars that he had dropped, more or less hung in the air and took minutes to fall to the floor. Gallia had essentially no gravity. There was no atmosphere, and one better not try to open the window, since the atmospheric pressure within the building was artificially maintained.

The gigantic sun indicated that Gallia was far inside the orbit of the earth. Herr Meyer suddenly saw another large disk rise above the horizon. Like the earth's moon, many dark areas could be seen, far more in fact. With his binoculars, Herr Meyer could see mountain ranges that formed craters, which were distributed just like the craters on the moon. The grazing sunlight caused their mountainous rims to cast long shadows over the lower-lying plains.

Then he heard a voice behind him. "Would you be able to tell it from the earth's moon if you did not know that it was Mercury?" Herr Meyer turned around and recognized the French scientist Palmiro Rosette, whose face was familiar from the illustrations in his edition of Verne's book.

"Is that Mercury then?" Obviously they must be very close to it, because it appeared about three times as large as the moon does from the earth.

"Take a look at Mercury's two different faces," said Rosette. "Like Janus it has two faces. One is saturated with innumerable impact craters, the other has few craters but great lava basins. Look at the screen."

Herr Meyer realized that he would be able to see pictures picked up by the telescope on the station's roof. The professor switched on the screen, and Herr Meyer saw a surface completely saturated with craters. Crater lay on top of crater (Figure 42).

Then came a valley that cut across the cratered landscape as if drawn with a straightedge. "On the left of the picture is the Caloris Basin." Herr Meyer saw a wide flat plain, only moderately pockmarked by craters.

"Some of the mountain scarps with steep slopes give the impression that some time in the past, Mercury shrank, causing mountain ranges to appear on its surface like wrinkles," the French scientist said. Elsewhere, Herr Meyer could see grooves and rift valleys that covered the surface like a spider's web.

"It actually looks like the moon," he decided. "Barren, apparently without atmosphere, and probably — like the moon — a stony, dusty, waterless waste." But the professor had already switched off the screen and did not reply.

Herr Meyer was reminded of the way Mercury rose over Gallia when he saw Venus in the milky sky. It must have been some weeks later, but Herr Meyer did not keep a diary in his dreams. The sun's disk had become noticeably smaller but was still larger than it is for us. Venus, which was just rising over the horizon, was larger than the disk of Mercury had been. When it had risen fully, it covered nearly half the sky. As the sun was setting on the opposite side of the sky, Venus was fully illuminated. Even with the naked eye, Herr Meyer could see differences from the appearance of Mercury. There was no sign of bright, sharp-edged spots on a dark background; here there was only a bright disk with fuzzy, hardly visible spots and some faint streaks (Figure 43).

"Nothing but clouds," said Rosette. "But I shouldn't sit around here. I want to turn on our radar, which will penetrate the clouds and show the true surface of Venus. After all, there's not much worth seeing here." But Herr Meyer could not tear his eyes away

Figure 42. *The cratered surface of Mercury resembles that of the moon. This picture was obtained by* Mariner 10 *on 29 March 1974 from a distance of about 30,000 kilometers. Roughly in the middle of the picture is a relatively young crater, 12 kilometers across, inside an older one several times its size (Photo: NASA/ JPL).*

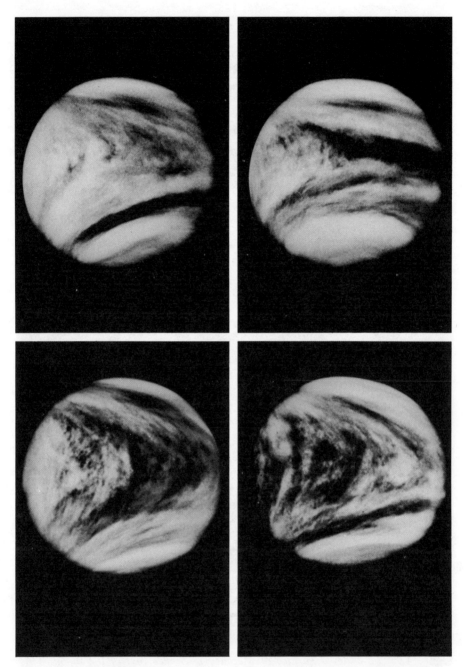

Figure 43. *These four, ultraviolet pictures of Venus, taken by the camera on board* Venus Orbiter *in February 1979, show the continually changing cloud structures that occur in the opaque atmosphere, which completely hides the surface of the planet (Photo: NASA/JPL).*

from the cloud-covered disk. "Clouds made of sulfuric-acid droplets cause sulfuric-acid rain," Rosette said, "but the rain evaporates into the atmosphere before it reaches the ground. The surface of Venus does not suffer from acid rain."

By this time a picture was visible on the screen. "Our radar is able to pierce those banks of clouds. Its pulses are reflected by the solid planetary surface, picked up again here, and turned into a picture." Professor Rosette was full of enthusiasm for his equipment.

"He must have got something wrong," thought Herr Meyer instinctively, because the picture was not the slightest bit similar to the featureless disk of Venus that he had seen through his binoculars. The screen showed mountain ranges separated by valleys, which often ran parallel to one another, broad plains, and here and there a crater (Figure 44).

"Those mountains have been caused by folding," thought Herr Meyer. At one point, he could see a large crater that was obviously very old, because a range of mountains ran right through the middle; it must have been formed after the crater.

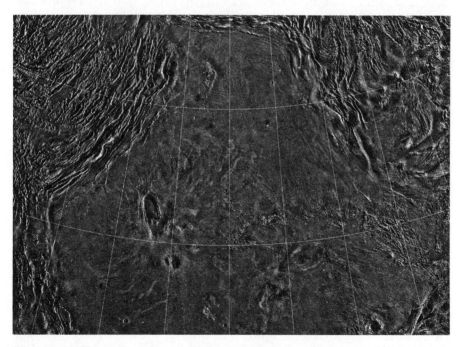

Figure 44. *This radar picture obtained through the cloud layers by the Soviet Venus probes* Venera 15 *and* 16 *shows mountain formations in the area of Lakshmi Regio, as well as features down to about 1 kilometer across. The method by which the picture was obtained is described in Appendix D.*

"There are very few craters on Venus," said Rosette, "because of its dense atmosphere. It protects the planet from meteorites, allowing only the very largest through, with all the smaller ones being burned up."

Herr Meyer looked at the screen for a long time and was rather surprised, because the mountains on Venus seemed to cast shadows, with one side of the mountain ranges being dark and the other bright. But direct sunlight could hardly penetrate the dense atmosphere, so what caused the shadows? Before he could solve the problem, he woke up.

■ A Radar Speed Trap Catches Mercury

Mercury takes 88 earth-days to orbit once around the sun. Its elliptical orbit departs farther from a circle than the orbit of any other planet, with the exception of the most distant, Pluto. During its orbit, Mercury is sometimes closer to, and sometimes farther away from the sun. The difference may amount to 24 million kilometers, which is about 41 percent of its average distance.

As an inner planet, Mercury crosses the earth–sun line twice in each orbit. By tradition, astronomers call these instants *conjunctions,* as we have already described for Venus in Figure 13. At superior conjunction, it is therefore on the opposite side of its orbit to the earth, behind the sun. At the next, inferior, conjunction, it is in front of the sun. At inferior conjunction, it is relatively close to the earth. However, it does not always transit the sun as seen from earth. Because the orbital planes of Mercury and Earth do not coincide, it nearly always fails to transit the sun's disk at inferior conjunction, just as it does not always disappear behind the sun at every superior conjunction. Venus also has inferior and superior conjunctions. The significance of the rare transits of Venus across the solar disk at inferior conjunction was described in Chapter 3.

As the innermost planet, Mercury is most strongly affected by the sun. Every square meter of its surface receives about seven times as much solar radiation as a similar area on Earth. As a result, it must be very hot on Mercury.

The sun causes tides even on the earth, so how much stronger its tidal effect must be on Mercury! Earlier in its history, when the planet was still fluid, tidal bulges must have braked its rotation over such a long period of time that by now the same side should always be turned toward the sun, just as the moon always turns the same side toward the earth. Until quite recently, astronomers pictured Mercury's rotation as follows: However fast it rotated originally, the tides caused by the sun must have braked the planet, so that it now has a "day side," which is always turned toward the sun, and a "night side," where the sun never rises. The day side captures all the radiation and must therefore be strongly heated, causing the surface of Mercury there to reach about 300°C, while the temperature of the night side must be about −200°C.

One can hardly expect to find an atmosphere, which would tend to equalize the temperatures, because at 4 kilometers per second, the escape velocity is closer to that of the moon than it is to that of the earth. In fact, so far, it has only been possible to detect tiny traces of atmospheric gases around Mercury.

But this picture of the temperatures on the day and night sides is not correct. The first evidence against it came in 1962 when radio astronomers attempted to measure the heat radiated by the planet. They found that the night side is significantly warmer than could be expected from the situation just described. The excess warmth of the night side of Mercury could be explained when it was assumed that it was rotating, so that every point on its surface was heated by the radiation from the sun. Then the day side, warmed by the sun, would later become the night side, and the nights would be warmer.

Confirmation came on 6 April 1965. As so often in astrophysics during the last few decades, radio astronomers were the ones who sprang the surprise. Rolf Dyce and Gordon Pettengill worked with Cornell University's radio telescope, which is located in Puerto Rico. Whereas the telescope is normally used to capture radio waves from space, the two researchers used the telescope's aerial as a transmitter, sending radio waves out into space. The wavelength was 70 centimeters—a wavelength used for radar. The two astronomers were using the radio telescope as a giant radar set. Just like the police who use their equipment to emit radar waves that are reflected by passing vehicles, so the astronomers sent radar waves out to Mercury and then captured the echo that it reflected back to Earth. Mercury was at inferior conjunction. But even then, it is so far from Earth that the radar signal, traveling at the speed of light, still requires five minutes to reach Mercury. It takes a further five minutes for the echo to return.

Details for the use of the reflected radar signal from a planet to determine its rotation are described in Appendix C, where we also explain why a radar signal, which is sent out at a sharply defined frequency, returns spread over a wider frequency range. From this broadening of a radar signal over a range of frequencies, it was possible to determine the rotational period of Mercury, with a surprising result: Mercury did not always turn the same side toward the sun.

As the moon orbits the earth, it rotates exactly once on its axis, thus always showing us the same side, but as Mercury completes one orbit around

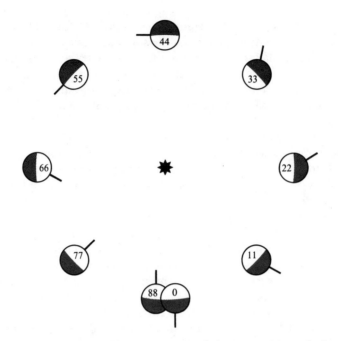

Figure 45. *As Mercury revolves once around the sun in 88 earth-days, it rotates one-and-a-half times on its axis. The numbers on the planet show the time in earth-days elapsed since position 0. A Mercurian year therefore amounts to half a Mercurian day. A flagpole on Mercury, drawn as a short line, is shown (bottom) at midnight at the start of a Mercurian day. After one orbit, it is midday. The flagpole will point to the same star another 58 days later, having taken 176 earth-days to go from midnight to midnight.*

the sun, it rotates one-and-a-half times around its axis. Two Mercurian years, therefore, equal one Mercurian day. Since the year on Mercury equals 88 earth-days, one Mercurian day last 176 earth-days (Figure 45).

▪ *The Craters of Mercury*

Actually, it should have been possible to detect the rotation of Mercury by looking through a telescope. The planet has light and dark patches like the moon, and it should have been possible to see how they move across the face of Mercury. But because of its proximity to the sun, Mercury is difficult to observe and is generally visible only shortly before sunrise or after sunset. But it is then so close to the horizon that its light has to pass through dense layers of the earth's atmosphere before arriving at the telescope. Optical observation of Mercury is difficult, and charts of Mercury drawn at the telescope were unreliable. Spots that people thought they saw were never seen again. Subsequently, but only after we had learned of Mercury's rotation from radar observations, it was found that the sparse visual observations did show the rotation of Mercury and its true period.

What are the markings on Mercury? The hour of truth came for Mercury when the first space probe was sent past it on 29 March 1974. The American space probe *Mariner 10* came within 700 kilometers of the surface and sent thousands of individual pictures back to Earth. Since then we know that the surface of Mercury is a cratered landscape similar to that of the moon. About 35 percent of the whole surface of the planet was mapped. Like the moon, Mercury is saturated with impact craters, and lava flows fill very extensive basins.

It seems that Mercury's "seas" of hardened lava preferentially occur on one side, like those on the moon. For the moon, we ascribe this to the effect of the earth. But what about Mercury, which travels through space all on its own, without any massive companion that would tend to intercept any wandering masses of rock and thus, to some extent, "protect" one side of Mercury?

What causes the asymmetry of the Mercurian surface? Was Mercury once a satellite of Venus? If it was, then perhaps tidal friction braked the rotation of Venus and slowly pushed Venus' moon farther away. (We discussed this for the earth–moon system in Chapter 3.) Finally, Venus' moon escaped from the planet's gravitational field and became a planet in its own right, whose asymmetrical surface carried evidence that it was once a satellite of a larger planet, which it orbited, always showing it the same side. But this is probably merely a nice idea.

■ *Mercury's Interior*

Although the *Mariner-10* space probe sent back a lot of information about the surface of Mercury, it was unable to see into its interior. The planet's mass has been known for a long time, because its gravity perturbs the other planets, particularly Venus, in their orbits around the sun, and thus, one can determine the mass of the perturbing body; it amounts to about four times the mass of the moon. Its diameter is not quite one-and-a-half times the diameter of the moon. From this, we deduce its mean density, which turns out to be 5 grams per cubic centimeter, which is closer to that of the earth (5.5 grams per cubic centimeter) than to that of the moon (3.3 grams per cubic centimeter). Mercury's surface may look much like the moon's, but its interior more closely resembles that of the earth. This is also suggested by its magnetic field, which is indeed much weaker than the earth's but does suggest that Mercury, like the earth, has a fluid core.

▪ Enshrouded Venus

When Galileo Galilei turned his telescope on the heavens and saw quite unsuspected objects, he maintained a bad habit that was common among scholars of his time. On the one hand, just like scientists today, they wanted to keep their discoveries to themselves for as long as possible, and on the other, they feared that someone else just as clever might make the same discovery and thus rob them of the credit of being first. The way out of this dilemma was to use an anagram. The result of a discovery was expressed as concisely as possible in a sentence—naturally in Latin—and then the letters were either scrambled at random or given in alphabetical order. Those who particularly favored this sort of game tried to arrange the letters so that the encrypted sequence had some other meaning. If someone later made the same discovery, it could be shown that the result had been expressed as an anagram long before. Credit for first discovery was assured, and the colleague had troubled himself for nothing. Galileo encrypted his discovery of Saturn's ring in this way (see Chapter 10) and sent the meaningless letters to Kepler, who tried in vain to decipher the anagram. Shortly afterwards, Galileo forwarded a second anagram to Kepler: "Haec immatura a me jam frustra leguntur oy," which was rather poor Latin for "These immature things I am searching for now in vain." What was behind this nonsensical sentence? Some time later, Galileo announced the solution: "Cynthiae figuras aemulatur mater amorum." If one knows that "the mother of love" means Venus and that Cynthia is the moon, one can understand the solution: "The mother of love imitates the figures of Cynthia." This was a rather flowery way of saying that through a telescope Venus showed phases like those of the moon. In Figure 3 we saw why an inferior planet, either Mercury or Venus, must exhibit phases.

When Venus was examined more closely through telescopes, people occasionally thought they saw darker areas, which they thought were markings on the rotating surface of Venus. They tried to determine the rotation of the planet from the appearance and disappearance of these markings. Reputable astronomers of the eighteenth and nineteenth centuries found that the rotation period of Venus seemed to be about 23 to 24 hours. In the third volume of his book *Kosmos,* Alexander von Humboldt cited an astronomer who stated that the rotation period was 23 hours, 21 minutes, and 21.93 seconds, that is, accurate to a hundredth of a second. In the history of astronomy, as in any other science, there have been many errors. I suspect, however, that rarely has a value given to such a degree of accuracy been so wrong. We now know that Venus rotates around its axis once in 244 earth-*days.*

The great William Herschel (see Chapter 11) and his son John, who followed in his father's footsteps, supported the view that the dark markings on Venus were cloud features, which could not be relied upon for determining the rotation period of the planet. That Venus was surrounded by a dense layer of clouds had been suspected from quite early on. Indeed, when the planet is close to inferior conjunction and therefore more or less between the sun and the earth, Venus shows a very narrow crescent, similar to the moon just before or after new moon. But unlike the moon, the two points of the crescent stretch far around the unilluminated portion, and the disk of Venus occasionally even appears surrounded by a ring of light. This is to be expected because, rather like a mirage, refraction in the planet's atmosphere allows light to spill over from the illuminated side onto the dark side.

Venus orbits the sun in 225 earth-days. Since its orbit is close to a circle, its distance from the sun is practically constant throughout a Venus-year. The planet is somewhat smaller than the earth and has about 81 percent of the earth's mass. Its density of 5.24 grams per cubic centimeter is similar to those of Mercury and the earth.

▪ *The Extraordinary Rotation of Venus*

On 14 December 1962, when the American space probe *Mariner-2* flew by another planet for the first time (at a distance of 35,000 kilometers), it confirmed that Venus has a dense cloud layer. Its infrared radiation suggested a temperature of 200°C. In 1964, Irvin Shapiro of the Massachusetts Institute of Technology succeeded in determining the rotation period of Venus. Again radar echoes were used. It was found that Venus rotated around its axis once in 244 days relative to the fixed stars. But unlike Mercury and all the other planets, it rotates *in the opposite direction* to its orbital revolution around the sun. We do not know the reason for this peculiarity, which is shown schematically in Figure 46.

Because radar signals penetrate the cloud layers around Venus and are reflected only by the solid surface, the measured rotation applies to the solid body of the planet but not to its atmosphere. The latter appears to rotate

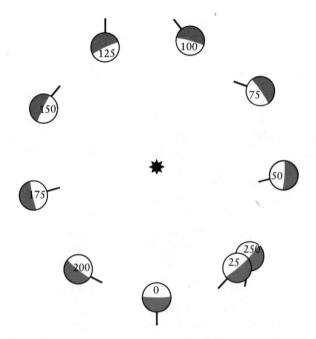

Figure 46. *The orbit of Venus, seen at a right angle to its plane, and from the direction in which its revolution appears* anticlockwise. *From this viewpoint, the planet rotates* clockwise *around its axis. The numbers shown on the planet, as in Figure 45, indicate the time in earth-days that has elapsed since position 0. If, at the beginning of a Venusian year—which amounts to 255 earth-days—a flagpole (bottom) points in the opposite direction to the sun, after 58 days it points toward the sun. About 117 days later, it again points toward the sun. On day 175, it is again midday. If, at the beginning of the orbit it was midnight for an observer standing near the flagpole, one revolution later there would still be some time until midnight, two days later. The Venus-year therefore amounts to somewhat less than two Venus-days.*

about 60 times as fast. It only takes about four days for the storm winds constantly blowing around the equator of Venus to transport the air masses once around the globe.

Moreover, for some time it seemed that our own earth played a part in both the extremely slow rate and in the unique direction of the rotation of Venus. Every 583.9 earth days, Venus and Earth closely approach one another as they orbit the sun. Venus is then at inferior conjunction, and the distance between the planets is only about 41 million kilometers. The rotation of Venus is such that at each inferior conjunction, the same side is turned toward us; however, recent measurements have not confirmed an exact relationship between the periods. There is a measurable difference. The earth actually has nothing to do with the rotation of Venus.

▪ *The Soviet Invasion of Venus*

Since 1961, Venus has been the preferred target for American and Soviet space probes. In February 1961, *Venera 1* was launched from the Soviet Union on a path toward Venus, approaching the planet to within 100,000 km. Since then, nothing more was heard from the probe. The American probe *Mariner 1* did not even get that far. After problems at the launch on 22 July 1962, the rocket had to be destroyed. But at the end of August, *Mariner 2* was launched and took pictures of Venus from a distance of 35,000 kilometers.

Then the Soviets returned to the fray. At the end of February 1966, *Venera 2* came within 24,000 kilometers of Venus. Unfortunately, the data relay back to Earth failed to function. But then came the milestone: On 1 March 1966, *Venera 3* soft-landed on the surface of Venus. Unfortunately this probe did not return any measurements, either. Finally, on 18 October 1967, *Venera 4* entered the atmosphere of Venus, and for 94 minutes, as it slowly descended by parachute, it radioed the temperature, pressure, and composition of the atmosphere back to earth. In 1970, Soviet scientists managed to soft-land another probe, *Venera 7,* on Venus. That probe lasted for 23 minutes and *Venera 8,* which landed two years later, returned measurements for 50 minutes. It is only thanks to later *Venera* probes that we have images of the surface of Venus.

In October 1975, two Soviet probes reached Venus, each consisting of two parts. While one was designed to drop through the atmosphere onto the planet's surface, the other would follow a circular orbit around Venus and relay the faint radio signals from the lander to earth. The project was a complete success. I shall shortly describe the conditions that prevailed beneath the opaque atmosphere of Venus.

Beta Regio on the surface of Venus was given its name at the time of the earliest radar measurements made from earth. (The technique of radar imaging is described in Appendix D.) The formations that were detected on the radar images of Venus were "Alpha," "Beta," and "Gamma," not exactly imaginative descriptions, but they have survived to this day. Since then, the mountains on Venus have been named after famous women, such as Colette and the Austrian physicist Lise Meitner, or mythological females, such as Leda, Diana, Niobe, Phoebe, and Rhea. After the *Challenger* accident in

January 1986, Soviet astronomers proposed that two craters should be given the names of the two women astronauts who were killed.

The mountain Rhea is a mighty 6000-meter peak not far from Beta Regio. The region is a rubble-strewn highland that slopes gently toward the east and is about 1000 meters above the average level of the surface. As usual, on 22 October 1975, the surface of Venus was illuminated by a fairly bright light that filtered down through the everlasting white layer of clouds covering the sky, where it was repeatedly scattered by tiny droplets of sulfuric acid. Below the lower haze layer, starting at an altitude of about 30 kilometers, the hurricane-force east–west current reached speeds of 40 meters per second, which amounts to 144 kilometers per hour. Given the high density of the atmosphere of Venus, even gentle winds exert a powerful force on anything that stands in their path.

It was on the eastern edge of Beta Regio, at a spot where the ground falls away sharply, that we got our first sight of the surface of Venus. Far above the cloud layer, a speeding capsule arrived. It entered the atmosphere of Venus at a shallow angle and at a speed of about 10 kilometers per second, compressing the gases in front of it. The gases became hot and also heated the foreign body arriving from space, but its heat shield absorbed the heat and protected the equipment inside it from being burned up. As if they wanted to prevent its entry, the gases in the atmosphere resisted the capsule and slowed it down. When it had penetrated the atmosphere's upper haze layer above the clouds, its velocity had dropped to 250 meters per second. The first parachute opened, and in a few seconds, the body lost even more of its dangerous speed. By then, it was inside the cloud layer of Venus. The heat shield had done its duty and was ejected. Three giant braking parachutes kept the remaining descent under control. Now the probe began the tasks that had been programmed into it months before. Its photocells examined the surrounding clouds, checking how much light was reflected when the clouds were illuminated by the probe, and determined that visibility was limited to a few kilometers. The probe dropped through clouds of sulfuric acid. The measurements showed that the atmosphere consisted mainly of carbon dioxide. By now the probe had dropped through the 20-kilometer-thick layer of clouds and had reached the lower haze layer. The descent could not take place too quickly, or else the probe would have smashed into the surface. But too slow a descent would also have been dangerous, because the temperature outside the capsule was 200°C. The equipment inside could not be kept cool and functioning for very long. The parachutes had to be released, because they were now slowing down the fall too much. The remainder of the atmosphere down to the surface of Venus still had to be investigated. At 05:13 Universal Time, the probe landed on the surface of Venus at about 7 meters per second, at longitude 293° west, and latitude 33° north, according to the coordinate system we use for Venus. The Soviet probe *Venera 9* had safely landed on a mountain slope at the edge of Beta

Regio. The measurements made during its descent had already been radioed to the mother craft orbiting Venus at a safe distance and had been relayed to Earth.

Now searchlights were turned on, and a camera took a wide-angle panoramic picture of the landscape. It was a dreary prospect: a field of boulders, consisting of loose, angular rocks, about half a meter across. The air pressure was 90 bar, which on Earth corresponds to the pressure at a depth of 900 meters below the sea. The outside temperature was 460°C. The probe resisted these conditions for nearly an hour and then fell silent.

Three days later, a second Soviet probe, *Venera 10*, landed about 2000 kilometers away. We have another panoramic picture of the surface of Venus from it, which again shows an inhospitable landscape full of boulders. After the first photographs of the surface of Venus, *Venera 13* and *Venera 14* sent back four more pictures in 1982.

The sharp-edged rocks seen by *Venera 9* gave cause for thought. Would not the violent storms on Venus have blasted away all the irregularities with wind-blown sand? Would not the acid vapors in the hot atmosphere at nearly 500°C have attacked the rocks long ago? Or were we dealing with relatively young material that had reached the surface from the interior just a short time before? What is the general state of volcanism on Venus? Are there volcanic mountains, volcanic craters, and regions filled with lava similar to the "maria" on the moon and on Mercury? The few photographs of Venus are unable to give us any information about that. A completely different technique reveals what the Soviet panoramic photographs could not show. American scientists first found that radar signals provide information not only about the rotation of a planet but also about the nature of its surface.

■ *The Americans Arrive*

In 1974, long after the first Soviet probes had soft-landed on Venus, the American probe *Mariner 10* flew relatively close to Venus, at a distance of less than one earth radius. But the Americans only really arrived with the *Pioneer-Venus* probes.

Pioneer-Venus 1 was launched on 20 May 1978, and on 8 August of the same year, *Pioneer-Venus 2* followed it on its way to our neighbor planet. Both space craft arrived there in December. Another probe, the *Sounder,* had already been released from the second spacecraft and was to plunge into the atmosphere of Venus. A few days later, three smaller probes known as *North, Day,* and *Night* were released. The mother craft continued its journey to Venus along with these four smaller probes. *Pioneer-Venus 1* was inserted as an artificial satellite into an orbit that followed an elongated elliptical path around Venus. With each orbit, it came to within 250 kilometers of the surface of Venus and then, 12 hours and 7 minutes later, it reached the opposite point in its orbit, about 66,000 kilometers from the planet. In the succeeding days, its orbit was frequently corrected from Earth, until it came within 150 kilometers of the surface of Venus.

On 9 December, the remnant of *Pioneer-Venus 2* and the four probes that had separated from it plunged into the atmosphere of Venus at various places. *Sounder* was slowed down by a parachute, but crashed onto the surface at about 36 kilometers per hour. The other three capsules had no parachutes; they were protected by heat shields but were solely intended to send back data during their 55-minute fall through the atmosphere. The *Day* probe, however, survived the impact and sent back data for another 67.5 minutes before it gave up the ghost. The data showed that the dust cloud that the probe sent up on impact took about three minutes to settle back onto the ground. Then the main body of *Pioneer-Venus 2* entered the atmosphere of Venus, but being without a heat shield, it burnt up about 100 kilometers above Themis Regio in the southern hemisphere of Venus. This brought the *Pioneer-Venus 2* mission to a close.

But *Pioneer-Venus 1* was still following its elliptical orbit around the planet, investigating the atmosphere and the underlying surface. The artificial satellite of Venus, which was now renamed *Venus-Orbiter,* determined that it was obviously impossible for hydrogen to be retained by the planet's gravity and that it would therefore escape from Venus. The planet is surrounded by a shell that is a few thousand kilometers thick and consists of hydrogen escaping into space. As Halley's Comet flew past Venus at a distance of 40 million kilometers in February 1986, the probe directed its instruments toward the comet and radioed its measurements back to earth. It confirmed that the comet was also surrounded by a cloud of hydrogen.

Radar equipment aboard *Venus-Orbiter* gave the first images of the mountain ranges on Venus. Although a camera could not pierce the layers of clouds around the planet, radar techniques were able to provide a view through Venus's outer cloak.

▪ *The Radar Picture of Venus*

Radar waves are extremely short radio waves of less than one meter (a typical medium-wave radio station transmits on 375 meters; a short-wave station on 49 meters.) For radar, mist and clouds are transparent, making it indispensable for modern sea and air transport. When radar waves encounter an object, they are reflected back into space in all directions. A fraction of the emitted radiation returns to its point of origin. We shall discuss in Appendix C how the rotation of planets can be determined from the earth by using radar. The method by which radar can obtain pictures of the surface of Venus through its extremely dense, surrounding layer of clouds is described in Appendix D.

The *Venus-Orbiter* carried radar equipment, and during its hundreds of orbits, it obtained close-up images of the surface of Venus and determined the height of the strip of land over which it was flying. Thus, we now know a lot about the surface, although only six true photographs of it — the *Venera* images — have been obtained.

Most of the surface of Venus consists of a widespread, hilly landscape with numerous impact craters. Two mountainous regions, or rather high plateaus, tower above the surface. On the equator, Aphrodite Terra (named after the Greek goddess of love) reaches heights of 4000 meters, while in the north, we find Ishtar Terra (named after the corresponding Babylonian goddess). In this mountainous area, the Maxwell peaks tower up to 10,800 meters. With all the other features on Venus being named after women, the Maxwell Montes, a group of extinct volcanoes, are the "odd man out": The name commemorates the English physicist, James Clerk Maxwell (1831 – 1879). Lower areas of the surface of Venus show few craterlike structures. They are probably basins that were filled with lava fairly recently, similar to the "maria" on the moon and Mercury.

The best radar images of the surface of Venus to date were provided by the Soviet probes *Venera 15* and *16* (Figure 44). Both were launched in June 1983 and reached Venus in October of the same year, entering elliptical orbits around the planet. They began to map Venus on 1 November 1983 (the method is described in Appendix D). By using radar wavelengths of 8 centimeters, they were able to detect features down to about 1 to 2 kilometers across from heights of 1000 to 2000 km. The *Magellan* probe is expected to map Venus with even better radar equipment. Its launch had to be delayed as a result of the *Challenger* catastrophe. On Friday, 5 May 1989, at 1:00 Universal Time, the astronauts Mark Lee and Mary Cleeve released the 3.4-ton spacecraft from the loading bay of the space shuttle *Atlantis*. As I write these lines, the spacecraft is on its way to our neighboring planet. In August 1990, after it has passed one-and-a-half times round the sun, it is due to be inserted into orbit around Venus. It will send radar images of the surface of Venus back to Earth, and it is hoped that these will show even more detail than the *Venera* images.

There are no seas on Venus, but it is quite possible that water once played an important role on the planet. At a later date, the water-vapor molecules were probably split into their constituent atoms, hydrogen and oxygen, in the upper layers of the atmosphere. Hydrogen escaped into space, and so over the course of time, water became scarcer and scarcer on Venus, until it finally more or less completely disappeared, leaving only a trace of water vapor in the atmosphere today.

▪ *The Venusian Greenhouse*

What causes the extremely high temperatures that the Venus probes could withstand for only about an hour? Venus is closer to the sun than the earth and thus intercepts more sunlight. But its clouds reflect a lot of the solar radiation, certainly more than the earth's atmosphere. The high temperatures on Venus are the result of a particular property of the gases in its atmosphere. Despite the strong reflection by the clouds, the energy arriving from the sun—which is primarily carried at visible wavelengths—does penetrate into the atmosphere of Venus and warms the planet's surface. The ground reradiates this energy in the form of infrared radiation. But the atmosphere of Venus consists primarily of carbon dioxide, a gas that is opaque to infrared radiation. The energy captured from the sun does not succeed in escaping and is stored, as in a greenhouse on earth.

Grassland warmed by energy captured from the sun radiates away some of the heat in the infrared region of the spectrum or gives it up to air flowing over it. It therefore attains only a moderate temperature. In a greenhouse, however, the closed glass windows prevent the escape of infrared radiation and the exchange of air. The temperature in a glasshouse is higher because the energy entering through the glass is trapped inside. On Venus, the energy arriving from the sun can only be radiated away with difficulty, because the atmosphere is barely transparent to infrared radiation. Therefore Venus is hot. It is the greenhouse effect that turns Venus into a fiery hell.

Markings on Mars

In six months a dozen small towns had been laid down upon the naked planet, filled with sizzling neon tubes and yellow electric bulbs. In all, some ninety thousand people came to Mars, and more, on Earth, were packing their grips. . . .

RAY BRADBURY
The Martian Chronicles (1946)

■

In H. G. Wells's classic science-fiction novel *The War of the Worlds,* the Martians were repulsive aggressors, who were on the verge of subjugating the whole world. No one was the slightest bit sorry when, one after the other, they fell victim to terrestrial diseases, with which their immune systems could not cope. According to Kurd Lasswitz, a German philosopher and one of the first science-fiction writers, Martians were noble beings, not only technically more advanced, but also with a higher level of ethics than man. Physically or genetically, however, they were hardly any different from us. A Martian, left behind by an expedition to earth, became the father of a son by an earth-woman; in addition, the son was semi-noble. So far, neither type of Martians have been discovered, because there appears to be no life at all on Mars.

▪ *Markings on Mars*

The Dutchman Christiaan Huygens (1629–1695) first noticed that, seen through a telescope, Mars had dark markings. With the improvements in telescopes, more and more features were discovered on the planet, which was already quite striking because of its reddish color. But even today with the very finest telescopes, the conditions under which Mars can be observed from the ground are far worse than those ever experienced by anyone wanting to recognize features on the moon with binoculars.

In the eighteenth century, it was discovered that the markings periodically changed their color; sometimes they were greenish and were therefore called "seas." Eventually, people recognized bright areas at the poles that alternately expanded and shrank, depending on which pole was tilted toward the sun as the planet moved around its orbit. We now know that the white polar caps fluctuate in accordance with the seasons. The northern one grows when it is winter in the northern hemisphere of Mars; the southern one follows half an orbit later. A Martian year amounts to 1.88 earth-years. As a planet outside the earth's orbit, it has a longer orbital period than the earth, as demanded by Kepler's third law (discussed in Chapter 1).

The orbit of Mars departs markedly from a circle, although not as strongly as that of Mercury. It was because of the orbit of Mars that Kepler realized that the planets move in ellipses and not in circles. When the earth has completed one orbit around the sun, Mars has traced out somewhat more than half of its orbit. Let us start with a time when the earth lies between the sun and Mars, at which time the earth and Mars are relatively close. Mars is then at opposition as seen from the earth (Figure 47; see also p. 57). After one earth-year, they are on approximately opposite sides of the sun, because Mars has moved around only part of its orbit in the meantime. After about 780 days, the earth has again caught up with Mars, resulting in the next opposition of Mars. But not all oppositions of Mars are equally close to the earth. Since neither the earth's orbit nor that of Mars is circular, Mars and the earth are not at the same distance from one another at each opposition. About every 15 years the oppositions fall where the orbits are particularly close. This is a field day for Mars observers, and one can expect to see more features on the Martian disk. During all favorable past oppositions of Mars, observers have pointed their telescopes at the red planet. Earth and Mars may come within 55 million kilometers of each other. But at unfavorable oppositions, the distance between them may be approximately twice that (Figure 47).

The dark markings change color over the course of the Martian seasons. The polar caps are brilliant, snow-white areas. Toward the end of the last century, it was concluded that the yellow markings that could be seen on Mars were obviously desert areas. Finally, it was noted that occasionally the

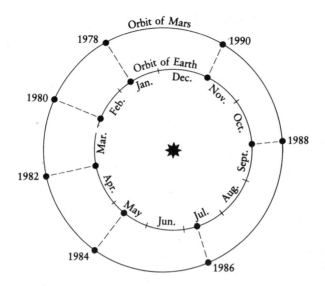

Figure 47. *The distance between the orbits of Earth and Mars varies at different oppositions. This diagram shows the positions of the two planets at some recent and future oppositions.*

disk of the planet appeared to be covered with a yellowish veil: These were dust storms that moved across the Martian landscape.

Ice and snow at the poles; thawing snow in the spring; occasionally greenish areas, which might be vegetation that changed with the season; storms that raged across sandy deserts; Mars became more and more earth-like. A sensational "discovery" came in 1877. The director of the Brera Observatory in Milan, Giovanni Virginio Schiaparelli (1835–1910), was a keen planetary observer. He observed Mercury and Venus, but mainly concentrated on Mars. Above all, he plotted the topography of the planet. From the way certain markings came and went, it had been known for a long time that Mars rotated. The length of the Martian day, which only exceeds that of the earth's by about 40 minutes, was known fairly accurately by the middle of the eighteenth century. Unlike the moon, which only allows us to see half of its surface, every region of the surface of Mars can be seen from Earth. While Schiaparelli tried to draw a map of Mars, he thought he had seen straight lines that ran across the bright areas between the dark markings. He called these features "canali," which can either mean a canal filled with water or a "channel." This was the origin of the "Martian canals," which inspired the human imagination. Because such geometrically regular features could not simply arise naturally, people came to believe that they must have been constructed by intelligent beings who irrigated the deserts of their planet, which was poorly endowed with water, via a vast system of canals.

▪ *The Search for the Inhabitants of Mars*

It was already known that the dark markings on Mars could not be seas because otherwise at oppositions of Mars, observers would have noticed sunlight reflected from the water surfaces. Water appeared to be available only in the polar caps in the form of ice. At the time of the thaw, dark zones seemed to appear around the edges of the polar caps. It was probably meltwater. Were the inhabitants of Mars attempting to channel this water across the whole surface of the planet in a system of canals? Not only Schiaparelli saw the canals; another observer saw a canal that ran right across a dark area. Had it been built through the middle of an area of vegetation? Was it not possible that the green areas owed their existence to the central canal, from which they were irrigated?

Although astronomers were relatively restrained about the idea of life on Mars, it had considerable appeal for amateurs and the general public. A rich businessman, Percival Lowell (1855–1916) of Boston, incurred the displeasure of many professional astronomers in America, when he promoted his views on intelligent beings on Mars with a lot of skill and considerable success. In 1894 he founded an observatory for the study of the planet in Flagstaff, Arizona. Today, the observatory is highly respected, with many important observations to its credit.

The more observers saw the canals, the more contradictory the whole matter became. Lowell saw them as being so extremely narrow that others wondered how he could detect such fine structures with his telescope. His assistant, meanwhile, saw much wider canals. Other observers were unable to detect any Martian canals at all, no matter how hard they tried. Lowell blamed this on the poor seeing conditions under which they had made their observations. His comment on the problems of properly viewing the canals was:

> When a fairly acute eyed observer sets himself to scan the telescopic disk of the planet in steady air, he will, . . . of a sudden become aware of a vision of a thread stretched across the orange areas. Gone as quickly as it came, he will instinctively doubt his own eyesight . . . [then] with the same startling abruptness, the thing stands before his eyes again.

Following Lowell's work, Eugene Antoniadi (1870–1944) devoted his attention to the Martian canals. After twenty years of intensive observation with the great refractor at Meudon near Paris (the telescope had an objective lens 81 centimeters in diameter), he concluded that the Martian canals did not exist. "The details anywhere on Mars show an extraordinarily irregular structure," he wrote. He actually managed to detect some fine detail in those areas where Schiaparelli and Lowell had seen canals. When viewed through a telescope with poor resolution, such markings can certainly appear to merge and resemble straight lines. What Schiaparelli saw as canals, Antoniadi recognized as consisting of a series of individual spots, certainly nothing with a canallike structure. The French planetary researcher Audoin Dollfus has probably shown this even more clearly. Where, under moderate seeing conditions, he saw canals like those of Schiaparelli and Lowell, under excellent conditions, he could detect individual, aligned markings, which were certainly not narrow canals that looked as if they had been drawn with a straightedge (Figure 48). Photographs taken from earth never showed any canals, but that was not surprising. Anyone who looks through a telescope using high magnification sees each object being examined as an image that is continuously in motion and fluctuating, just as the landscape appears to shimmer when we look at it across an asphalt road heated by the sun. It drives home the fact that we are at the bottom of an agitated sea of air. The eye is able to recognize some features for an instant before they are changed by the movement of the image; but a photographic camera records the continually changing image on a plate over a long period of time. The result is a fuzzy picture that is incapable of showing small features.

Martian canals went out of fashion in the 1920s. Even if there were no canal builders on Mars, people did not completely give up the idea of plant life on the red planet. Even in the sixties, respected authors still claimed that

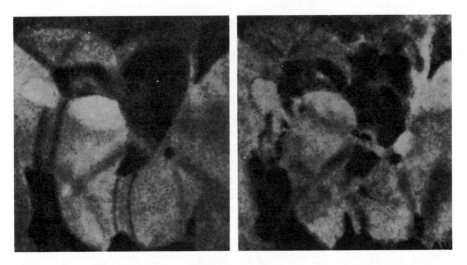

Figure 48. *The Martian canals in drawings of the same region of Mars by Audoin Dollfus, made under poor (left) and good (right) seeing conditions. Whereas the left-hand picture shows the Martian canals, the picture with the higher resolution on the right shows how the irregular surface features trick the eye into seeing Martian canals and other apparently regular features.*

lichens, mosses, and algae might be able to exist in the Martian climate. The fact that the dark markings appeared again after every dust storm also attracted attention. Because of its numerous dust storms, Mars should have long ago become covered in an even layer of dust. Did the fact that the markings reappeared in their old form not mean that the areas of vegetation were repeatedly renewed, and that they grew over any newly spread layers of dust? People were not prepared to abandon the idea of life on Mars easily.

Herr Meyer on a Collision Course with a Martian Moon

He did not exactly know how long he had been flying through space. Was it two weeks or two months ago that they had passed the earth's orbit? Herr Meyer remembered how he had seen the earth from far out in space, but still near enough for him to see with the naked eye the blue of the oceans in between the white clouds, the white polar caps, and the landmasses. He had been able to recognize Africa easily. Then his gaze had turned to the pale moon, which seemed to hang motionless close to the earth, but which suddenly and visibly darkened and appeared dim and red in the darkness. Herr Meyer had observed the beginning of a total eclipse of the moon from Comet Gallia, on which he was again traveling through the solar system.

But that was a long time ago. Herr Meyer had lost sight of the earth and the moon, and the sun had become noticeably smaller. Gallia was now apparently far outside the earth's orbit.

Then a large, red disk rose over the horizon. At first only one of the Martian poles could be seen. Herr Meyer recognized the white polar cap, which almost dazzled him with its light. The more the disk rose over Gallia's horizon, the more of the large, red surface he could see. Mars was not evenly illuminated by the sun, because Gallia had approached it from a direction from which the boundary between day and night could be seen (Color Plate 1).

"It is now dawn over the Tharsis region," said Professor Rosette as Herr Meyer looked at the boundary between day and night. Three mountains were emerging alongside one another into the day-light. The low elevation of the sun caused them to cast long shadows, which enabled Herr Meyer to appreciate their heights. He could see whitish patches on the nightside of the largest of them.

"Those are Ascraeus Mons, Pavonis Mons, and Arsia Mons," said Professor Rosette. "Clouds of water ice generally cover the peaks of the higher mountains as they move from night into day. It will soon be dawn on Olympus Mons. It is far higher than anything that we are used to on earth. It rises nearly 30 kilometers above the average level of the surface of Mars. Look at the image of Mars on the screen. You can see things that no astronomer can explain. There are volcanoes, like Olympus Mons. In that case, it is easy to see how lava, erupted from its summit, has flowed down its sides and then hardened. You can see fine streaks and wrinkles that show the direction in which the material has flowed. You can also see impact craters that have been partly flooded by lava from neighboring volcanoes. But look for yourself." And he pointed to the screen, which was showing the image of Mars that was being picked up by the telescope on the roof. Olympus Mons was now in the middle of the picture, and the clouds of ice crystals seemed to be slowly fading away. Then the view moved across the Martian landscape. Herr Meyer suddenly saw a remarkable shape in the middle of the irregular landscape that was pock-marked with craters. At first sight it reminded him of some fossil in a layer of slate. Looking at it more closely, it was obviously an enormous rift valley (see Color Plate 2).

"This system of canyons stretches for about 4000 kilometers along the Martian equator," explained the professor. "But you can see for yourself what other curiosities there are on Mars." He altered the picture visible on the screen, and Herr Meyer now saw a system of rivers. Small streams arose more or less anywhere, joining onto larger and larger streams, until they formed a wide river, which, along with others, disappeared in the distance (Figure 49).

"That Martian landscape has been formed by vast floods," Herr Meyer exclaimed. "Yes, but without a single drop of water. There is scarcely any water there. It is unable to feed any rivers. The only water on the surface of Mars is in the polar caps." As he said this, Rosette changed the picture on the screen again.

"The northern polar cap of Mars," he explained, but Herr Meyer did not only see a bright expanse of ice, seen from above and filling the screen; he also noticed dark lines snaking and spiralling outward from the planet's pole. And when the telescope and the camera moved away from the icy polar cap, Herr Meyer could see a broad expanse of sand dunes (Figure 50).

"The wind on Mars has a very strong influence on the landforms, even though the atmosphere is very thin. You can see how the wind flows around many craters and how the wind's irregularities have distributed the sand.

"Look," the French professor continued, busying himself with his computer keyboard, "while the polar region is quiet, there's a major

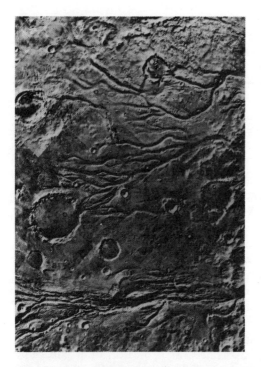

Figure 49. *These formations on the surface of Mars give the impression of being fully developed river courses. The image is a mosaic assembled from various Viking Orbiter 1 images. Undoubtedly, the channels were created at some time in the past by a flowing liquid. At a few points, the channels cut across impact craters, but in other places the impacts occurred later and are superimposed on the river beds. The temperatures on Mars have always been such that it could hardly ever have been possible for large amounts of water to exist in a fluid state. Thus, these channels on Mars remain a puzzle to planetary researchers. Were they once filled with water? (Photo: NASA/JPL.)*

Figure 50. *A gigantic field of dunes near the northern polar cap of Mars. At the point indicated by the arrow, the dunes probably cover an old crater (Photo: NASA/JPL).*

sandstorm raging near the equator." The image moved across the planet and came to a halt at a yellowish area. Actually, there was nothing to be seen: All features had completely disappeared.

"All you are looking at now are clouds of dust, which prevent you from seeing anything of the surface of the planet." But Herr Meyer had noticed that Rosette kept looking at the clock and then peering out into the blackness of the sky.

"I am waiting for Deimos," the professor said. "It ought to be here any minute." Just then, a bright star rose over the horizon of Gallia.

"The Martian moon Deimos is a very irregular body with an average diameter of about 12 kilometers. We are looking at it from a distance of 14,000 kilometers, so it is not surprising that to the eye, it only appears as a tiny disk. We can see quite a bit more in the telescope: It has a few craters but is otherwise very uninteresting. It orbits its planet once every one-and-a-quarter earth-days. Gallia is now crossing the orbit of the other Martian moon, Phobos. If my calculations are correct, we shall pass dangerously close to it." Herr Meyer scanned the horizon, but could see nothing. Suddenly, a giant potato seemed to fill the sky. He had no idea where this enormous, irregular body had come from. It was growing visibly, and within a few minutes Herr Meyer could see that the object, which was half illuminated by the sun and half in shadow, was saturated with craters. The closer Phobos came, the more details could be seen even with the naked eye (Figure 51).

There were not only craters but also grooves and scratches, as if at some time the elongated body had squeezed through a narrow gap. Herr Meyer noticed that Professor Rosette was becoming visibly more and more nervous. Spellbound, he was gazing out of the window at the gigantic rock that was getting nearer and nearer. By chance, Herr Meyer glanced at the television screen, which was showing the telescope's image of the Martian moon. He realized that Phobos was not heading straight for Gallia and that the two bodies were not on a collision course, because he saw that some of the grooves at the edge of the picture were slowly moving out of the field of view. Herr Meyer realized that Rosette had still not noticed this. As he called his attention to the screen, the motion of the Martian moon became even more noticeable. Gallia was not about to collide with Phobos. The French professor smiled with relief.

"If it had come to a collision, we would have come off worst. Phobos has an average diameter of 22 kilometers, whereas Gallia is only about 5 kilometers across. Phobos would have just added another impact crater, but Gallia would have been smashed to pieces by the impact. It looks like Phobos has already survived even more powerful collisions in the past."

Figure 51. *The surface of the Martian moon Phobos exhibits impact craters and remarkable parallel "scratches." Irregularly shaped Phobos is about 20 kilometers across at its narrowest point. Its major diameter is about 27 kilometers (Photo: NASA/JPL).*

▪ A Visit to Mars

Numerous science-fiction novels describe how Martians have attacked the earth. In 1962, it was time to hit back. The counterattack by the Soviet probe *Mars 1*, which began its flight to the red planet on 1 November 1962, was not very convincing. After a few weeks and after having traveled about 100 million kilometers, radio contact with it was lost and nothing more was heard from it. Two years later, the Americans launched *Mariner 3* toward Mars, but that also failed because a protective shield failed to release at the right moment. The first success came a short time later with *Mariner 4*, which flew past Mars in July 1965 at a distance of about 10,000 kilometers, and gave us the first close-up pictures of the planet. The transmission of the 22 images was a long-winded process. In order to ensure that the faint signal from the probe could be picked up successfully, it was necessary to transmit the signal slowly, picture element by picture element. But even that was not possible immediately, because after taking the 22d picture, the probe passed behind Mars, where it remained hidden for nearly 11 hours. The South African tracking station near Johannesburg was the first to be able to receive the pictures again. The transmission rate was so slow that each picture took 8 hours and 40 minutes. Then the magnetic tape on the space probe was rewound and the transmission repeated in order to correct the signals that were distorted in the first transmission. This was how we also obtained the first close-up pictures of Venus. (The Soviet pictures of this planet were not to be available for another 10 years.)

The areas of Mars photographed lay in a single strip that crossed the side of planet that was turned toward the sun. The first image showed the horizon and a patchy area. One of the Martian canals, the Hades canal, should have crossed the field that was photographed. The image showed nothing. Only in the third image were any details visible. The photograph covered an area of 350 by 500 kilometers and showed that there were craters on Mars! This fact was even clearer on the 7th and 14th images. There were craters with small peaks in their centers, and craters on top of larger craters, which therefore must have been formed one after the other. There were grooves like those known on the surface of the moon. Mars had a lunar landscape. But none of the pictures showed any Martian canals.

Soviet and American probes have visited the planet many times since then. They have not always been as successful as *Mariner 4*. Radio contact was lost with the Soviet probe *Zond 2*, but then, in 1969, the American *Mariner 6* and 7 sent back a total of 201 images of Mars from a distance of only about 3200 kilometers. In 1974, the Soviets succeeded in placing their probe *Mars 5* into orbit around the planet. After many failures on the part of both space agencies, two American probes, *Viking 1* and *2*, finally soft-landed on Mars in 1976, sent back pictures from the surface of the planet, and carried out experiments on the Martian soil (Color Plate 3). At the same time, after having released the landers, the two mother crafts, mapped Mars from orbit, providing Martian research with many exciting pictures.

▪ *The Tragedy of the Soviet* Phobos *Mission*

Somehow the Soviets were not as lucky with Mars as they were with Venus. Their *Mars 3* did soft-land on the surface in 1971, but contact was lost after only four minutes. Unfortunately, the Soviets also had dreadful luck with their last Mars project. Years of hard work by groups of researchers from 13 countries were suddenly lost when the *Phobos* catastrophes occurred in 1988 and 1989.

In July 1988, two rockets were launched from the Soviet space center of Baikonur on the Aral Sea, sending two nearly identical space probes on their way to Mars. *Phobos 1* and *Phobos 2* were both intended to investigate the Martian moon Phobos, hence their names. It soon became clear how sensible it was to send two probes, both equipped with more or less the same instruments, when *Phobos 1* was lost in the autumn of 1988 due to a stupid error. During its flight to the red planet, the instruments were provided with electricity obtained from solar cells. In order to capture enough solar energy, the two surfaces fitted with these cells had to face the sun. The cells charged the batteries, from which the instruments drew their electricity. Because of a faulty command sent to *Phobos 1,* the orientation of the probe was changed, and the cells were no longer exposed to sunlight. Before the error could be corrected, the instruments on board had drained the batteries. Without electricity, the probe could no longer receive and carry out any commands from earth.

But the sister probe was still under control. During its flight, it returned data on electrically charged particles ejected from the sun. In December, the transmitter on *Phobos 2* failed, and a backup transmitter had to be switched on. The probe was inserted into orbit around Mars on 29 January 1989. During the following weeks, it orbited the planet, found radiation belts, and slowly approached the Martian moon Phobos. On 1 April, the probe was programmed to fly past Phobos at a distance of only 50 meters and to release a piece of equipment that would land on it. On 27 March, the probe, which had previously taken photographs of the Martian surface, was turned toward the satellite to determine its orbit more accurately and to provide data for the fly-by maneuver. The signal indicating that the turn had been completed reached the earth. That was the last that was heard of *Phobos 2*. It is assumed that either the on-board computer or the transmitter failed. Which of the two pieces of equipment was responsible for wiping out years of work by all the researchers who had entrusted their instruments to the *Phobos* probes, we shall never know. The ones who got off most lightly were Professor Susan McKenna-Lawlor and her co-workers from Ireland. At the last moment, the Irish were able to detect the Martian radiation belts, in which—as in the Earth's radiation belts—electrically charged particles are trapped.

It is assumed that the Soviet Union is still very concerned with Mars and is making attempts to have men living in a space station for as long as possible in preparation for a manned flight to Mars.

▪ *A Bleak Landscape*

The craters did not all result from impacts. Olympus Mons is just one of the Martian volcanoes that occur in mountainous areas and that appear to have been formed by uplift in the past. Olympus Mons itself seems to have been created only a few hundred million years ago, because its flanks show only a few impact craters. Thus, it cannot have been exposed to cosmic bombardment for long. The impact crater Yuty has a diameter of 18 kilometers and shows remnants of mud flows that appear to have been ejected during the impact. Another impact crater at the edge of Yuty is obviously older, because the mud flow has nearly filled it to the brim.

Mars does not just have an eventful volcanic history; even today it does not appear to have quieted down completely. The Martian surface exhibits occasional fractures that cut across the landscape in straight lines. Like the geologically active folds and rift valleys on the earth's surface, they were probably formed by slow motion within the interior of the planet, on which the plates that form the outer crust float.

The polar caps of Mars partly consist of frozen carbon dioxide, like our dry ice, and partly of water ice. There are certainly no glaciers several meters thick. During each hemisphere's thaw in the spring, the polar caps melt so rapidly that the snow line retreats by several kilometers each day. They appear to be rather more like hoarfrost.

▪ *The Martian Rivers*

It would therefore appear that not much water is stored in the polar caps. The two *Viking* landers came to rest some 6500 kilometers apart. But the general view was the same: rocks, sand dunes, and boring boulder fields. The dreary, bleak Martian landscape must once have been far more convivial, however. The *Viking 1* orbiter, which photographed Mars from orbit in 1976, sent back images of dried-up river systems (Figure 49). Today, although there is no flowing water on the surface of Mars, it shows features that can only have been produced by some fluid. It is not only river systems that can be seen. There are craters whose walls have been breached at one point; features on the ground indicate that some liquid has flowed out of them. What is particularly remarkable is that the flow must have persisted for a long time, because the openings in the crater walls have been so thoroughly eroded. One gets the impression that over thousands of years, rainwater in a crater lake has flowed out into a river. The dried-up riverbeds have wide, trough-shaped cross sections. Teardrop-shaped "islands" frequently occur within them. It is very difficult to reconcile the existence of rivers in the past with the current dryness of the planet.

Where is the Martian water now? It is suspected that there are layers of permanently frozen water—permafrost—beneath the surface of Mars. We do not know what caused the change on Mars that allowed its water to vanish. One solution is perhaps offered by the theory that after the formation of the planet, the surface was so heavily bombarded by meteorites that it became porous to a depth of several kilometers. The water previously existing on Mars might have sunk into these fractured layers, where it is now locked in the form of eternal ice. The ice would melt only immediately beneath the surface of Mars at the height of summer. In fact, a yearly cycle has been observed in the radar reflectivity of the Martian surface. Radar waves are strongly affected by the existence of any liquid. A re-examination (in the mid-1980s) of 2000 images returned by the *Viking* and *Mariner* probes suggested that there is more water on and beneath the surface of Mars than had previously been supposed.

▪ A Very Thin Atmosphere

The atmospheric pressure on Mars is less than one one-hundredth of the air pressure at sea level on earth. On earth, the air pressure at a height of about four times that of Mt. Everest is comparable to that at the surface of Mars. Carbon dioxide is the principal component of the Martian atmosphere, followed by nitrogen and the inert gas argon. Oxygen is only present in trace amounts, and there is also very little water. If all the water in the Martian atmosphere were condensed, it would form a layer of only one two-hundredth of a millimeter over the whole surface of the planet.

When the *Viking* probes landed in Mars's northern hemisphere, it was summer. The midday temperatures were around −30°C. In winter, the temperatures are around −140°C. Although the Martian atmosphere is so thin, clouds do form in it. They often consist of crystals of frozen carbon dioxide. Winds of up to 200 kilometers per hour lift grains of sand and dust into the air. Such hurricane-force winds occur mainly in the southern hemisphere when the planet's orbit brings it closest to the sun and solar heating is at its greatest. The *Viking 1* lander experienced such a dust storm and took photographs of the boulder-strewn landscape before, during, and after the storm, which did not appear to do much damage.

The Martian mountains represent barriers to such dust storms. Dark streaks on the surface, stretching downwind from many mountain peaks like banners, probably indicate where eddies have formed on the lee sides of the peaks and swept away the dust and sand, exposing the dark underlying surface. Elsewhere, features like sand dunes show the power of the wind to affect the shape of the surface.

There are good reasons for believing that Mars once had a much thicker atmosphere. There are two forms of the element nitrogen, which do not differ chemically (see Note 1 to Appendix B), and Mars has much more of the heavier nitrogen 15 relative to the lighter nitrogen 14 than the earth does. If we assume that originally Mars had the same proportions of the two types of nitrogen as the earth, then we must assume that Mars has lost much of its lighter nitrogen 14. We have already seen in Chapter 3 that a planet loses lighter atoms faster than heavier ones. If we assume that nitrogen 15 remained behind on Mars while nitrogen 14 escaped, then the atmosphere must probably have been ten times as thick.

Did air—now only present in amounts that would not allow human beings to exist—and water—which now shows only traces, but once probably existed in liquid form—allow life on Mars?

▪ *No Signs of Life*

By 29 July, 1976, the *Viking 1* lander had been on the surface of Mars for eight days. The first pictures, including the first color pictures, had already been sent back to earth. Now a sampling arm was extended from the probe and carefully reached out into the rubble field. The first sample of soil was obtained. This was the beginning of the experiments carried out on Mars by the *Viking* landers. The soil samples were chemically and physically investigated inside the probes and were found to contain materials, primarily silicon, that we find in terrestrial soils. When compared with terrestrial rocks, the Martian soil contains relatively more iron. Different processes must have occurred when Mars was formed than took place on Earth. We now believe that nearly all the iron sank to the center of our planet, so that the earth's core now consists of nickel and iron. Instead, on Mars, iron appears to have remained close to the surface.

It is now clear that the reddish color of the planet is caused by oxidized iron — in other words, just ordinary rust. The red planet is the rusty planet, as had been suggested more than 50 years ago by the astronomer Rupert Wildt (1905–1976) of Göttingen.

The samples of Martian soil were not only subjected to a geological analysis. The more burning — and to the general public a more interesting — question concerned the possibility of life on Mars.

How could a robot standing in the midst of a boulder field and the dunes on Mars recognize whether there was any life present? The images of the landscape transmitted directly from the probes showed no signs of life. There were no Martians waving at the camera; there were no trees or plants to be seen; no animal jumped into the field of view; and there were no remains of buildings to show the existence of a vanished civilization. Might there be at least traces of primitive life, such as single-celled organisms, bacteria, or viruses?

The search was carried out only for life that resembled the sole lifeform with which we are familiar, the terrestrial one. This means that attempts were only made to identify life based on the chemical compounds of carbon. There are good reasons for this, since these compounds offer the best possibility for the development of organized, self-replicating structures. Such forms of life should take in carbon dioxide from the air, like our plants, in order to breathe, and would therefore enrich the surrounding atmosphere with substances that were not there previously. In other words, they would have to ingest food and excrete various substances into their environment. The potential nutrient materials for Martian microbes provided in the test chambers were labeled, in that they contained radioactive carbon, that is, carbon 14, rather than ordinary carbon. However, the Martian microbes touched neither the carbon-dioxide atmosphere nor any other nutrients that were offered them.

Even when the sampling arm of the *Viking* 2 lander pushed aside one of the stones on the ground, under which microscopic inhabitants of Mars might possibly have been sheltering from the deadly effects of the sun's ultraviolet radiation, no signs of life as we understand it could be seen.

But the search for life was carried out at only two sites. A dried-up river bed would have probably been more suitable and might at least have shown signs of past life, but one would have to devise other tests to detect it. In any case, so far we have found not the slightest indication of life on Mars. This was not altered in the slightest by the remarkable rock, photographed by the *Viking 1* lander, which seems to have the letter "B" scratched on its surface, nor by a hill photographed from orbit by the *Viking 1* orbiter, that looked like a human head, with hair, eyes, mouth, and nose (Figure 52). It was just as if an artist had modeled it on the surface of Mars — an accidental natural

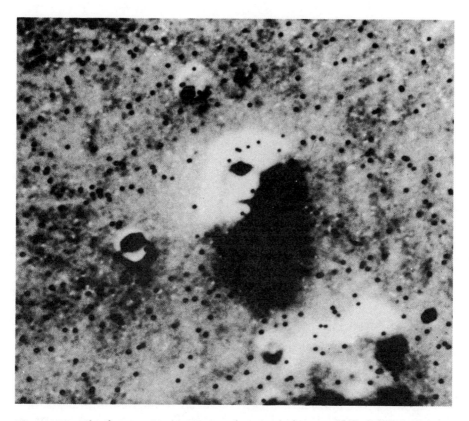

Figure 52. *The face on Mars. During the search for a suitable landing site for Viking 2 in 1976, this photograph was taken of a hill some 500 meters wide and 750 meters long that is illuminated by sunlight to give the impression of a human face. This impression is strongly reinforced by a black spot that forms one nostril, but which, like all the other black spots in the image, was caused by transmission errors (Photo: NASA/JPL).*

formation that under particular lighting conditions looked just like a human face.

▪ *A Race to Mars?*

Will Soviet cosmonauts be the first to reach Mars? From time to time, it is rumored that the Soviet Union is preparing to send a manned mission to Mars. That strengthens the position of those in the United States who support a manned flight to Mars. Optimists estimate the cost at 17 billion U.S. dollars. The NASA program manager, Jesco von Puttkammer, maintains that 50 to 60 billion dollars is a more realistic figure.

In fact, one would have to begin with a starting weight of about 900 tons to transport a crew, together with supplies and fuel, as far as the red planet, carry out a landing, and then return. In the foreseeable future, there is no rocket that could lift that weight from earth. Even the great *Saturn V*, which carried the *Apollo* astronauts to the moon, could lift only an eighth of that weight, and the space shuttle only about one-thirtieth. The material would have to be carried up to a space station in installments and assembled there.

Optimists estimate that the first manned base on Mars will be established between the years 2010 and 2020. It is hoped to set up bases on the two Martian moons even earlier. There are even studies in progress to establish how the materials to be found on Mars can be used in the life-support system for the members of the expedition. As we have seen, Mars is not a place where this is particularly easy: The carbon-dioxide atmosphere cannot support human life. In addition, there is the chronic lack of water.

But plants can thrive in carbon dioxide, giving off oxygen, and sunlight for the plants is available. So one can imagine using greenhouses to provide food and oxygen for the members of the expedition. Such a self-sustaining system requires water above all else. It is hoped that ice can be found beneath the Martian surface. A fine "report" on the colonization of Mars—half science-fiction and half a popular scientific study—has been written by the science journalist Michael Allaby and the chemist James Lovelock.[2]

Plans for a base on Mars are, however, likely to interfere with plans for a base on the moon (as discussed in Chapter 4). The former would cost 50 to 60 billion dollars, and the latter is estimated at 52 billion. In comparison, the cost of the entire *Apollo* program in the 1960s, with its high point of landing men on the moon, would nowadays come to 80 billion dollars. Certainly the decision will have to be taken as to which project should go ahead.

A by-product of a manned flight to Mars would be obtaining soil samples from the planet. Given the high cost of the mission, each individual Martian rock would be very expensive. Yet, eight rocks have so far been found on Earth that are suspected of having originated on Mars, and which have not cost us a penny.

When rocks from space crash into planets or their satellites, material is blasted upward. If the velocity of the ejected rocks is greater than the escape velocity, then they do not fall back to the surface. We saw at the end of Chapter 4 that this can happen with the moon. With Venus it is unlikely, because the ejected rocks would probably be vaporized in its dense atmosphere. The escape velocity of Mars is 5 kilometers per second, less than half the value for the earth, but double that of the moon. The thin Martian atmosphere presents no real obstacle to the ejected material, however. As with the moon, material from Mars can be flung out into space through impacts.

Might it be possible for such a rock to be intercepted by the earth? It is estimated that the earth captures about one hundred times more material from the moon than it does from Mars. The eight meteorites in question differ chemically from terrestrial and lunar material. They differ even more from other meteorites. Gases trapped within them resemble the Martian atmosphere as sampled by the various space probes.

Only a very small fraction of the rocks ejected from Mars fall onto the earth. Most mingle with the numerous small bodies that can be found in the space between the planets. These are the subject of the next chapter.

SEVEN

Comets

October 10. The comet in this evening's examination represented an extraordinary phenomenon. The brush, fan or gleam of light . . . was clearly perceptible, issuing from the nucleus, which was now 17 arc seconds in diameter and shooting into the coma; the glances at times being very strong, and of a different aspect from the other parts of the luminosity.

ADMIRAL W. H. SMYTH on Halley's Comet (1835)

■

The major planets show a certain regularity; their orbits are nearly all very close to a circle and lie in approximately the same plane. They all revolve around the sun in the same direction, and most of them rotate around their individual axes in exactly the same direction. But there are numerous bodies in the solar system that are not nearly as well behaved.

The most striking of these are the comets. They come from the farthest reaches of the planetary system on very elongated elliptical orbits that bring them right in toward the sun and thus close to the earth. The brightest of them may attract attention because of their tails, which frequently stretch far across the sky (Figure 53).

What sort of bodies are these, that come from the depths of space and were regarded in the past with foreboding and fear? The American astronomer Fred Whipple has broken the spell by giving a very prosaic answer: they

Figure 53. *A photograph of Comet Mrkos in 1957. Its gas tail, which points in the direction opposite to the sun, is straight, its shape being controlled by the solar wind. The broad dust tail to its left is slightly curved (Photo: Palomar Observatory).*

are dirty snowballs. Indeed, modern observations confirm that comets are kilometer-sized blocks of ice, within which dust and rocks are trapped, and which arrive in the inner regions of our planetary system from the cold depths of space.

Comet Gallia

For weeks, Herr Meyer had looked forward to the television program that would show direct pictures taken by the *Giotto* space probe from the immediate vicinity of Halley's Comet. The program was disappointing, however. A pop group sang a song about Halley, a bartender described a drink he had concocted in honor of the comet, and a hairdresser did the same for a cometary hairstyle. In between, various people were interviewed. There was a fortune-teller and a woman who had seen flying saucers. An astronomer tried to explain the comet, but for Herr Meyer the boring professor had nothing new to say. So he dozed off and found himself at the base station on Comet Gallia again. Professor Rosette was standing beside him at the window. Herr Meyer looked out at the dark landscape.

"I have to go down the shaft today and collect some soil samples. Do you want to come with me?" asked Professor Rosette. Herr Meyer nodded, curious what it would be like to take such an excursion across Gallia under almost weightless conditions.

A short while later, they were outside the base. They had put on heated space suits and had hooked themselves onto a cable that was lying on the surface.

"Such precautions are essential," explained Rosette. "The escape velocity here is only four millimeters per second. We are practically weightless. The slightest movement would send us into space, hence the precautions. All the buildings and the cable that we are hooked onto are anchored deep in the ice." Herr Meyer looked around; he did not see any ice. Rosette seemed to notice his surprise.

"We are standing on the surface of Gallia's nucleus," he said. "It is about 10 kilometers long, but only about 5 kilometers across. In the middle, it is even narrower, and sooner or later it will break into two pieces. The *cometary nucleus* consists of a mixture of water-ice, other frozen gases, and rocks of various sizes, down to the finest grains of dust, all of which are trapped in the ice. When the comet comes close to the sun, the ice in its nucleus vaporizes and streams away into space, carrying the dust and smaller rocks along with it. The larger fragments remain behind. They form a black crust, which protects the underlying ice from the sun."

By now, they had moved several hundred meters along the cable. The sun was high above in the milky-white sky. Then Herr Meyer saw a light on the horizon. Like a cloud, it was standing motionless over the horizon, but reached right down to the ground. The light seemed to originate from it. Then Herr Meyer saw that it was a broad shaft of light, which started at the horizon and rose almost as far as the zenith. Herr Meyer stopped.

"The dust particles break free after dawn," the professor explained. "Between the layers of rubble that remain, there are always areas where the warmth of the sun can reach and melt the underlying ice. The steam blasts the dust, and even some of the smaller particles, out into space. Because of the very low escape velocity, gas, dust, and anything else that is blown off escape from the comet completely. Its nucleus loses more and more mass. A gas and dust shell, some tens of thousands of kilometers across, surrounds the nucleus; this forms the *head* of a comet. The jets of gas and dust occur only on the sunward side. At sunset, the jets decline, and become active again at sunrise. The jet that you are now looking at is the strongest that has developed on Gallia. I have been observing it for years. At its base, the sun has already excavated a very deep trench into the nucleus of our comet. Only when we are outside the orbit of Jupiter in about half an earth-year's time, will Gallia stop degassing. Then Gallia's sky will be perfectly black."

Now Herr Meyer understood why the sky on Gallia was never completely dark. They were in the middle of the head of the comet, and the light from the sun illuminated the gas and dust cloud and was scattered onto the night side. But there was something in the professor's explanation that he did not understand.

"If gas and dust are being released only on Gallia's day side, the jets of gas must be directed toward the sun. But aren't comet tails always turned away from the sun?" he asked.

"They initially point towards the sun but then, at a certain distance from the nucleus, they are bent backward by the pressure of solar radiation and by the solar wind. Once farther out, they point in the opposite direction to the sun."

During this conversation, they had continued to work their way along the cable and had come to a deep pit in which there were various huts.

"Now we have to go down the shaft," said Rosette. When he saw the inquiring look on Herr Meyer's face through the window of his helmet, he added: "We have driven a shaft right through Gallia, so that we can investigate the interior of the comet."

They had now carefully unhooked themselves from the cable and hooked onto lines running between the huts, which they were approaching. A lot of people in space suits were working with various machines that Herr Meyer did not recognize. Then he saw a circular hole in the ground. It had a diameter of about one meter and was lined with some synthetic material. A cable, wound on a drum, ran over several rollers down into the opening.

"We need to get right to the center of Gallia. So we have to hook ourselves onto the cable and let ourselves be pulled a few kilometers down into the shaft."

"Couldn't we just simply jump into the hole?" asked Herr Meyer. "With the very low gravity, we would fall very slowly."

"It would take too long," replied the professor. "It would take us hours to reach the center."

The trip using the cable, which unwound from the drum and seemed to be wound in on the other side of Gallia, was very comfortable. The walls of the shaft were only lined at the top;

farther down, walls of ice slid by, glistening in the light of Rosette's hand lamp. The cable came to a halt when they reached the center of Gallia. Rosette began to collect samples from the walls and put them into containers that he had brought with him.

"It is water ice, together with frozen carbon monoxide and carbon dioxide, and also some ammonia and methane."

Even the feeble gravity had completely disappeared. "Because there is almost exactly as much mass in any direction outward from this point, the forces of attraction cancel each other out, and there is a complete state of weightlessness here. It would be exactly the same at the center of the earth," the professor explained.

They remained inside the shaft for several hours. Then, thanks to the cable, which was now being pulled in the other direction, they soon found themselves back at the hut. On the way back to the base, Rosette showed Herr Meyer the comet's tail. The sun had set by now, and the gas jet on the horizon had petered out. But in the sky, there was a bright spot about twice the width of a man's hand in diameter. From the center of it, a narrow, bright streak pointed off to the right. Herr Meyer was disappointed. He knew what comets looked like when seen from the earth. Now that he was in the immediate vicinity of one, he had expected to see an absolutely wonderful spectacle. What was that bright spot in the sky? He looked at Rosette questioningly.

"Gallia's tail is pointing straight out into space," the professor said. "But we are standing on the nucleus, where it begins. So, the comet's tail points directly away from us, and we are looking along the tail, rather than looking at it from the side. Perhaps I ought to be more precise: We are looking along Gallia's gas tail, which points straight out into space, but we see it as a round patch on the sky."

"And what's that streak pointing to the right?" asked Herr Meyer.

"That's the dust. Whereas the gas is blown away from the sun in a straight line, the dust moves differently, although it also moves away from the sun. It is affected by the pressure of sunlight, but the outward pressure of the light is not much greater than the force of the sun's gravity acting on the dust. So, the tiny particles move away from the comet on curved paths. The dust tails of comets are curved. Looking out from the cometary nucleus, we can see it partly from one side. Therefore, it looks like a streak of light, more like the usual view of a comet as seen from earth." By the time they had finished this conversation, they were back at the base.

When Herr Meyer woke up, his television set was still on, but the program on the comet had finished. Herr Meyer had slept entirely through the direct transmission of pictures from Halley's Comet.

▪ *The Invincible Armada Heading for Halley*

The first craft had already been under way for several years, when the other five were launched. In December 1984, the Soviet Union launched the two, one-ton *Vega 1* and *Vega 2* probes at an interval of just a few days. The first two letters of their names indicate their first target: Venus. In June 1985, they flew past the planet, ejected two landers and dropped two balloon-mounted instrument capsules into the atmosphere. Then the two parent craft devoted their attention to the second target, indicated by the second half of their names: "Ga" are the first two letters of "Gallei," the Russian way of writing "Halley." The spacecraft carried various instruments on toward Halley's Comet, which they would reach in March 1986. In January 1985, the Japanese probe MS-T5 was launched. It passed no closer than 7 million kilometers to the comet, whereas Japan's *Planet-A,* launched on 14 August, came within 200,000 kilometers of the comet. On 2 July 1985, the Europeans launched their *Giotto* probe, named after Giotto di Bondone (1267–1337), the Florentine painter. In his renowned fresco in the Scrovegni Chapel in Padua, he had painted the Adoration of the Magi, taking Halley's Comet as it appeared in the sky in 1301 as a model for the star of Bethlehem. Of all the probes, *Giotto* passed closest to the comet.

But years before any of these five space probes were sent on their way toward Halley's Comet, the *ISEE-3* probe (see Chapter 2) had been launched, without anyone suspecting that it would eventually be sent hunting a comet. Ever since it had been flung out into space by the earth's moon for its encounter in September 1985 with Comet Giacobini–Zinner (see below), it had been called *ICE* for "International Comet Explorer." In October 1985, after its initial cometary mission, it continued to hunt for comets. In March 1986, it was positioned directly between the sun and Halley's Comet and was able to make measurements of the solar wind before it reached the celebrated comet.

What is Halley's Comet, the target of this whole fleet of spacecraft? And what are comets in general?

▪ *The Comet of 1577*

In the year 1577, the tail of a comet stretched far across the sky; its head shone brightly. At the end of November, it was possible to see a second, smaller tail that projected from one side of the head. The comet was so bright that it could occasionally be detected even in daylight. People anxiously stared at the sky, maintaining that the comet would bring war, famine, and the plague, and that its emanations would poison the water in the streams and the grass in the pastures, causing the death of both man and beast.

But the comet did not only stimulate the imagination of the general public. The Danish astronomer Tycho Brahe had it on his observational program and followed the path of the comet across the sky for two-and-a-half months. Never before had a comet been tracked so accurately. Tycho wanted to find out once and for all what comets really were.

This question had various answers. Unlike other heavenly bodies, comets appeared suddenly and unpredictably in the sky. Within a few weeks, they moved across the constellations, and disappeared again. Their tails, looking like a trail of smoke coming from the head, sometimes stretched far across the sky. Were such objects part of the realm of the planets, or were they phenomena occurring in the earth's atmosphere, like clouds and rainbows? Was Aristotle right, when he said that comets were terrestrial objects, vapors that were carried up into the upper regions of the atmosphere, where they caught fire and slowly burnt away, leaving a trail of smoke behind them? If so, then they would not be true celestial objects. This was, in fact, the reason why reputable Greek astronomers, such as Hipparchos and Ptolemy, never mentioned comets in their works. Or was Seneca, Nero's teacher, right when he said that comets belonged to the planetary sphere?

Tycho Brahe's measurements of the comet of 1577 resolved the issue. For the first time, it was possible to deduce from them how far comets were from the earth. The basic idea was simple: If a comet follows a regular path across the sky and is observed from earth, its motion does not appear completely regular. Because astronomical observers are situated on the rotating earth, they are carried in a circle around its axis once a day. As a result, at any given time, they see the comet from a slightly different point. The position of the comet relative to the background stars changes because of the earth's rotation (Figure 54).

We mentioned this phenomenon, the so-called diurnal parallax, in Chapter 3, and showed how it was used in attempts to determine the Earth–Mars distance. It was known in principle long before Tycho Brahe's time. It had been used to determine the distance of the moon, because its motion on the sky clearly shows a daily variation. Now the same method ought to show how far the comet was from the earth. In his series of measurements, Tycho Brahe searched for signs of a diurnal parallax, but he found nothing. This negative result meant that the path of the comet was so far from the earth that its diurnal parallax was smaller than Tycho Brahe could measure. He therefore concluded that the comet was much farther away than the moon. This was the first step to modern cometary research. Aristotle had been wrong, and Seneca had been correct: Comets belong to the realm of the planets.

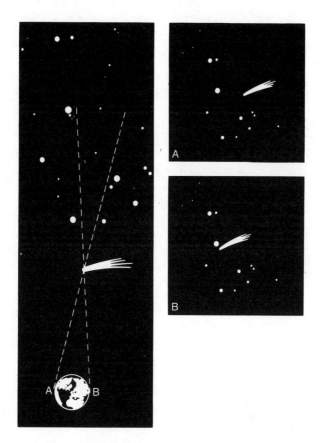

Figure 54. *To an observer rotating with the earth, a comet appears projected against different points on the background of fixed stars. The view from points A and B (12 hours later) are shown schematically on the right. The effect is known as diurnal parallax, which was discussed in Figure 25.*

▪ *Parabola or Ellipse?*

The next chapter in cometary research was written about 120 years later. In the meantime, Galileo, Kepler, and Newton had discovered the laws governing planetary motions. It was known that the planets moved in elliptical orbits around the sun, and it had been recognized that celestial bodies could also move in parabolic or hyperbolic orbits in the sun's gravitational field (see Figure 16). Whereas the planets move in elliptical orbits around the sun, comets appeared to move in parabolic paths, coming from infinity, entering our solar system, and then disappearing again into the depths of space.

Newton had shown how to determine the position of the comet's orbit from its motion across the sky; thus, it was now possible to calculate the orbit of any well-observed comet.

Edmond Halley, Newton's friend, about whom we talked earlier, tried to determine from the series of observations of earlier comets how their parabolic orbits were orientated in space. When he determined the parabolic orbit of the bright comet of 1682, it occurred to him that it was very similar to the orbit of the comet observed by Johannes Kepler in 1607. And a comet in 1531 observed by the astronomer Apianus—we shall hear more of him later—had moved in practically the same orbit. This gave Halley the idea that these three appearances might be one and the same comet. But then the orbit could not be a parabola, because celestial objects moving on parabolic orbits would never return! It must be a very elongated ellipse. Celestial bodies moving on elliptical orbits return at regular intervals to perihelion (the point on their orbit nearest to the sun). This was just what the appearance of comets in 1531, 1607, and 1682 indicated.

The comet did not appear to return at precisely regular intervals, because the period varied between 75 and 76 years. Halley had a suspicion that this irregularity was the result of perturbations that the gravitational fields of the major planets exerted on the motion of the comet—a suspicion that was later shown to be perfectly correct. If Halley's theory about the regular return of the comet was the right one, then the comet must continue to return at regular intervals. So Halley ventured to predict the next return as being in 1758. In fact, the comet returned at the end of 1758—16 years after Halley's death. Since then, it has reappeared on three occasions: 1835, 1910, and 1986. (Although it is by no means the brightest comet, it is the best known. But that is the way it goes: Those who are the most popular are not always the brightest.) It was last at its most distant point from the sun in 1948, after which it approached the sun again, being met by the fleet of cometary probes in the mid 1980s.

Halley had shown that the comet belonged to our solar system and that its orbit never took it away from the sun's gravitational field. Nowadays, we know of many comets that return at regular intervals. It also seems that no comets actually enter the solar system from outside and leave it after a single passage. All of them appear to belong to our solar system, even if many of them have such elongated orbits that the farthest points of their paths lie far outside the orbits of the most distant known planets. Figure 78 (in Chapter 11) shows the orbit of Halley's Comet, together with those of the earth and of the outer planets.

The discovery of the properties of cometary orbits took place in the sixteenth and seventeenth centuries, whereas the most important steps in discovering the nature of comets were taken only in the last two centuries. After Newton's work, it was possible to determine the position of cometary orbits in space, and this led to an approximate idea of the spatial extent of the phenomena that they presented. What was surprising was the enormous length of the cometary tails, which often stretched over a hundred million kilometers and more, a distance that approaches that between the earth and the sun.

▪ *A Comet Breaks Up*

How much material is actually present in these vast objects? It is always difficult to determine the mass of a celestial body. Despite this it proved possible to ascertain the amount of mass gathered in the head of a comet — to weigh a comet, so to speak. In 1770, a comet came so close to Jupiter that its head passed between the satellites of Jupiter as they orbited the planet. The gravitational attraction of the cometary mass should actually have perturbed the orbits of the satellites. But no irregularities in their motions could be detected. From this, it could be concluded that the mass of the head of the comet was very small. When another comet came close to the earth, the latter's orbit around the sun was not affected, or at least the effect was not measurable. In 1805, the French mathematician Pierre Simon Laplace (1749 – 1827) concluded from this that the comet must have had less than one millionth of the mass of the earth.

The Austrian captain Wilhelm von Biela (1782 – 1856) discovered a comet on 27 February 1826. In accordance with the custom of naming a comet after its discoverer, this comet became known as Biela's Comet. Because of its extraordinary behavior, it brought considerable fame to Biela. He had determined the comet's orbital period as being 6¾ years, but he was not responsible for the fact that the comet split in two, more or less in front of the astronomers' very eyes, in December 1845. The two fragments continued alongside one another in their orbits around the sun. At the next return, in 1852, both parts were seen again. Since then, there has been no trace of Comet Biela. It seems to have completely disappeared. After its breakup had occurred, the gravitational attraction between the two parts mutually affected one another. This mutual perturbation allowed their masses to be estimated; this again showed that a comet has a very low mass. We now know that a comet contains perhaps a trillion tons of material, which is less than one billionth of the mass of the earth.

When we look at a picture of a comet, such as that shown in Figure 53, we get the impression that as the comet moves, it leaves its tail behind it. This impression is quite incorrect. As early as the first half of the sixteenth century, the German mathematics teacher Apianus (1495 – 1552)[1] had discovered that however comets move, their tails are always turned away from the sun. A comet never flies through space trailing its tail behind it in its orbit (Figure 55).

At its apparition in 1910, when Halley's Comet crossed the line joining the earth and the sun on the morning of 19 May, it passed in front of the disk of the sun, as seen from earth. It should have been possible to see the nucleus of the comet as a dark spot against the disk of the sun. But despite careful observation, nothing was seen. The nucleus was too small and the head of the comet too transparent. Because the tail was pointing away from the sun at that moment, our planet actually moved through the comet's tail.

Once again, a comet gave rise to alarm. Might not the gases in the comet's tail poison the earth's atmosphere? Nothing happened, however. We now know that the material in cometary tails is so thin that the few molecules that do penetrate into the atmosphere during such an encounter have absolutely no effect—at least not in comparison with the pollution that we create ourselves.

In addition to their gas tails, many comets have a second tail that consists of tiny solid particles. Their diameter is of the order of one thousandth of a millimeter, so we are talking of the very finest dust. Although the gas tail points more or less exactly in the direction opposite to the sun, the dust tail normally curves and points in a different direction (Figure 53).

The essential part of any comet is its nucleus. It is the source of the gas and dust particles that form the comet's tails. When a comet is far from the sun, it has no tail. It is only when it is close to the sun that gas and dust are released from the nucleus by the action of solar radiation to form the tails.

Figure 55. *The tail of a comet always points away from the sun, irrespective of the direction in which it is moving. The tail therefore appears to precede the comet when it is moving away from the sun.*

Color Plate 1 *Mars at dawn, taken in August 1976 by* Viking *Orbiter. North is at the top. In the picture, the surface of the planet is rotating from left to right, that is, from the night side to the day side. In the upper half of the picture, Ascraeus Mons is illuminated. Water-ice clouds stretch out from its peak toward the west. In the southern hemisphere (bottom) frost surrounds the Argyre Basin (Photo: JPL/NASA).*

Color Plate 2 *Mars as seen from* Viking Orbiter. *Below the center is the dark horizontal strip of Valles Marineris, to its left near the border are the volcanoes Ascraeus (top) and Pavonis Mons and Arsia Mons (below). Above Valles Marineris one can see the plain Lunae Planum (Photo: JPL/NASA).*

Color Plate 3 *A Martian landscape. The view from the* Viking I *Lander looks southward over the Chryse Planitia to the horizon, which is about 3 kilometers from the Lander.*

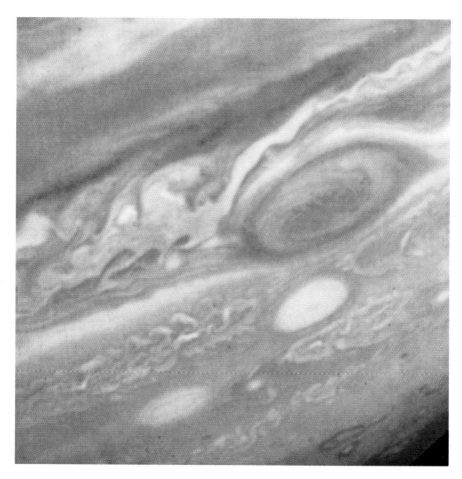

Color Plate 4 *The Great Red Spot (GRS) and its surroundings, taken by Voyager 2 from a distance of 5 million kilometers. South is toward the bottom. The atmosphere, which continually forms eddies, is streaming past the GRS. South of it a white oval is visible. The GRS is a whirlwind that has lasted for at least a hundred years. The smallest features visible on this picture have a diameter of 95 kilometers (Photo: JPL/NASA).*

Color Plate 5 *Io, the innermost of the moons of Jupiter to be discovered by Galileo. This picture was made by combining many individual images taken by Voyager 1 on 4 March 1979 from a distance of 862,000 kilometers. The ring-shaped object with the dark center in the middle of the picture is an active volcano. Io has many active volcanoes, which may eject material up to heights of 280 kilometers during eruptions. Io's diameter of 3600 kilometers is slightly larger than that of the moon. Its colors make it very striking: orange, black, and white predominate. Featureless, smooth plains are found between the volcanoes. The volcanic activity continuously brings sulfur-rich material to the surface (Photo: JPL/NASA).*

Color Plate 6 *Saturn and its rings as seen by* Voyager 1 *in November 1980 (Photo: JPL/NASA).*

Color Plate 7 *Neptune, photographed by Voyager 2, exhibits a dense atmosphere in which a dark spot indicates where a circular storm has formed. White clouds move at high speeds around the planet. Faint belts resemble the structure of belts parallel to the equator that is found on Jupiter (Photo: JPL/NASA).*

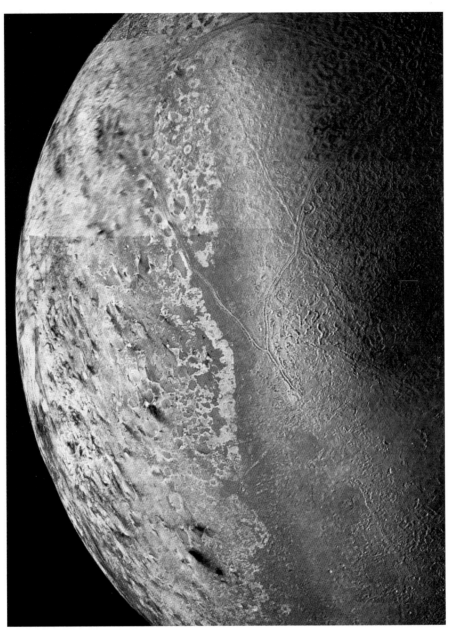

Color Plate 8 *Neptune's satellite Triton has few impact craters on its wrinkled surface, which is furrowed by long grooves. Particularly noticeable is the sharp transition between the bright areas, here illuminated by the sun, with many individual features and contours, and the even plains, crossed only by a few valleys, which in this image appear in the twilight zone (Photo: JPL/NASA).*

The material in the gas tail moves away from the sun at a high velocity. Velocities of around 100 kilometers per second have been observed.

▪ *Cometary Tails in the Solar Wind*

Because the tail of a comet always points away from the sun and because the gas in the tail moves outwards at such high velocities, the suspicion arose very early on that there must be some force acting on the atoms of the gas, and counteracting the force of gravity. The particles in the dust tail also appeared to experience a repulsive force, but one that was much smaller than that acting on the gas. The gravitational attraction of the sun forced the dust particles into curved paths, but gravity appeared to have no affect on the gas molecules.

The question of the repulsive force caused by the sun concerned astronomers for a long time. The only force known, and that could therefore be considered, was the pressure exerted by light from the sun on the particles in cometary dust tails. The pressure of light is so minute that we never notice it in everyday life. However, in free space, it is sufficient to push small, solid particles, and also atoms and molecules, away from the sun, against the force of gravity. We now know that it is responsible for the motion of the particles in dust tails. In 1943, however, the astronomer Karl Wurm (1899–1975) was able to show that radiation pressure is far too weak to account for the high velocities observed in gas tails.

But the high velocities had been observed and therefore required an explanation. Since the gas always moved away from the sun, the cause had to be sought in the sun. Around 1950, this gave the astrophysicist Ludwig Biermann (1907–1986)[2] the idea that perhaps the solar system was pervaded by a continuous stream of particles from the sun, which dragged the molecules from cometary nuclei along with them. It was known that in occasional eruptions on the sun, clouds of gas were ejected into space (they are, for example, responsible for the aurora borealis on the earth). However, Biermann now suggested that, quite apart from the eruptions, the sun emitted a steady "wind" of electrically charged particles. These are primarily protons —in other words, hydrogen nuclei—and electrons. These particles dragged the electrically charged fraction of the gas released from a cometary nucleus along with them, while the neutral molecules remained in the head of the comet. The solar wind, predicted by Biermann to account for the behavior of cometary tails, has since been detected by space probes and its strength and direction have been measured. So this answers the question of why cometary tails always point away from the sun.

Biermann also had the idea of producing an artificial cloud of gas in space in order to study the effect of the solar wind streaming past it. This would enable us to find out something about the mutual effects between cometary gases and the solar wind. The "artificial-comet" experiment was carried out several times.[3] The series culminated on 27 December 1984. On that date, at 12:32 Universal Time, two canisters carrying a total of 1.25 kilograms of barium were exploded at a distance of about 110,000 kilometers from the earth, creating a cloud of barium vapor. Initially, it glowed green. Then sunlight stripped electrons from the barium atoms, giving rise to barium ions. The glow turned violet. Then the electrically charged particles in the solar wind captured the newly created ions of barium and swept them far away into space. Within a few minutes, a cometary tail several thousand kilometers long had been created.

▪ The Evolution of Comets

Let us try to summarize the sequence of events for a typical comet. Its life begins out in the depths of space, at the edge of the solar system. Up to 10 trillion tons of material are contained in a body with a maximum diameter of a few tens of kilometers, the "nucleus" of a comet. These materials are dust and condensed gases, probably ice and frozen carbon dioxide, in which both the minute dust particles and some larger pieces of solid material are embedded. At that distance from the sun, the comet moves at less than walking pace over stretches of its orbit that are measured in trillions of kilometers. It has existed for a long period of time far outside the orbits of the planets, probably so far outside that in the course of millions of years, it has been repeatedly perturbed by the gravity of passing stars.

But the gravitational attraction of the sun is the strongest force acting upon it. It slowly responds to it and moves inward, toward the sun. After thousands of years, it reaches the orbit of Pluto, the outermost planet. The closer it comes to the sun, the faster it moves. When it reaches the orbit of Jupiter, the light and particle radiation from the sun begin to affect its surface. The frozen gases vaporize and form a gaseous shell some 10,000 to 100,000 kilometers in diameter around the nucleus, creating the "head" of the comet. To astronomers on the earth, the object appears as a fuzzy patch, which can be distinguished from the other nebulous objects on the sky by its motion.

The trapped, solid fraction is released along with the gases. The dust moves under the influence of both gravity and radiation pressure, forming the dust tail. While the electrically neutral molecules of gas form a shell surrounding the cometary head, the electrically charged molecules are captured by the solar wind that is streaming past. They are whipped away to form the gas tail. The larger solid particles that are released from the slowly evaporating nucleus remain within the head. Under the influence of various perturbing forces, they will later move away from the nucleus.

When the comet reaches the point of its orbit closest to the sun, gas production is greatest. Its tail stretches far across the solar system and may even reach a length of as much as 100 million kilometers. The comet has now become very bright and preoccupies the press, the public, astronomers, and astrologers on earth.

Then the amount of radiation received from the sun declines, and gas production diminishes. As the comet moves outward again, it slowly reverts to the frozen, inactive body that it was originally. In total, it may have lost about one one-thousandth of its mass during its visit to the interior of the solar system. Then, for tens, hundreds, or even perhaps millions of years, depending on the size of its orbit, it remains far from the sun. But eventually, it turns around, and the game is repeated.

Not all comets have a long life. On 30 August 1979, a comet fell into the sun. By chance, an American military satellite photographed the event. Before it disappeared into the sun, the comet had a gas and dust tail nearly five million kilometers long. It crashed into the surface of the sun at a speed of about 300 kilometers per second.

■ Colds from a Comet's Tail

The English astrophysicist Fred Hoyle, who is responsible for many advances in modern astronomy, is of the opinion that there are living bacteria in comets. He even claims that the earth is repeatedly infected with new cosmic germs and that probably many virus illnesses that suddenly appear actually come from space. I do not intend to discuss the long controversy that he aroused with these suggestions. It has been known for a long time that there are organic molecules in comets. But organic molecules are not life. Compounds that primarily consist of carbon and hydrogen atoms are called organic; however, they are not always associated with life. For example, molecules of formic acid — which on earth is produced by ants — are found in the molecular clouds between the stars. But just because of the existence of this acid, we cannot conclude that ants live out there between the stars.

Hoyle bases his theory of bacteria in comets on the results of a measurement that has been made. When a cosmic dust cloud lies in front of a star, it transmits varying amounts of radiation of different colors. Like a filter, it blocks certain colors but transmits others more or less without alteration. Dusts from different materials exhibit different amounts of absorption. The absorption properties of cosmic dust do actually resemble those of bacteria.

The procedure for determining the chemical composition of cosmic dust particles resembles the spectral analysis of gases, whose atoms remove specific wavelengths from any light passing through them. In gases, the light is removed over very narrow ranges of wavelength. Light from the sun is absorbed at thousands of different wavelengths in its atmosphere. Because each chemical element is responsible for specific wavelengths, the gases in the solar atmosphere can be analyzed chemically, without having a sample in a test tube. Each atom leaves a specific fingerprint on the light, enabling it to be identified. Unfortunately, this only applies to gaseous materials. A dust cloud absorbs only a very broad, fuzzy band that is a mixture of colors. Identification of particular types of material is rather ambiguous. Although the absorption characteristics of a *gas* identify it as certainly as a fingerprint identifies a criminal, an astronomer wanting to identify *solid materials* from just their absorption is more like a detective trying to identify a criminal from a composite sketch. There are very few astrophysicists who agree with Hoyle's theory of bacteria in cosmic dust.

▪ Flight through a Comet's Tail

After its long wait lasting several years at one of the sun/earth Lagrangian points (see Chapter 2), and after playing billiards with the moon and the earth, the probe *ICE* (see Chapter 2) was sent on its way to encounter Comet Giacobini–Zinner. On 11 September 1985, it passed less than 8000 kilometers from the comet. On board were 13 different experiments. The probe's original task was to investigate the solar wind. Apart from one French and one German experiment, all the others had been designed by American researchers. Now some of the experiments were to be used to investigate the vicinity of a comet.

At first, the experiments measured only the wind of particles coming from the sun. At 9:10 Universal Time, the probe crossed the comet's "bow shock," that is, the surface around a comet where the solar wind and the cometary wind collide. In this thin sheet, the particle density rises from its very low value in the solar wind to the relatively high value that prevails near the comet. Then *ICE* flew through a region where some of the material originated in the sun and some in the cometary nucleus. It soon encountered the cometary tail proper. The experiments detected molecules of water and of carbon monoxide. At the same time, the radio receiver registered extremely strong radio emission. Three hours and twelve minutes later, the probe reached the bow shock on the other side of the comet, having cut right across the comet's tail. Just as it had originally passed from a region of low-particle density to one where particle density was high, the value now dropped again, and the probe was back in the region of the thin solar wind.

During the three hours it spent inside the bow shock, the probe was hit by about 100 particles of dust, whose size was roughly just a few thousandths of a millimeter. None of these particles that crashed into the probe, which was flying through the tail with a velocity of 21 kilometers per second, did any damage. When its expedition to a comet is compared with the conditions experienced by *Giotto* half a year later, one could even say that *ICE* had operated under idyllic conditions.

▪ *The Night of the Comet*

On the night of 13–14 March 1986, the European space probe *Giotto* was due to fly by Halley's Comet.

Preparations had started long before. On 2 July 1985, an *Ariane* rocket placed the 750-kilogram probe into earth orbit. *Giotto* was then placed on a path that was to take it to Halley's Comet by the following March, closer to a comet than any other man-made object. Eleven research groups from Switzerland, the United Kingdom, Ireland, France, and the German Federal Republic had instruments on board. The most spectacular was the color camera developed by Ludwig Biermann's student, Horst Uwe Keller, and his team from the Max Planck Institute for Aeronomy. This was to obtain close-up pictures of the comet and radio them to earth, because it was planned to send *Giotto* past the nucleus of the comet at a distance of only 500 kilometers. The probe therefore had to pass right through the middle of the comet's head.

Giotto was moving parallel to the earth's orbit, but closer to the sun and therefore faster. The probe spun around its axis once every four seconds. In order to ensure that radio contact was not broken, the aerial had to be kept constantly pointed at the earth. Once the probe started to tumble, it would no longer be possible to control it. In the event of this occurring, a form of emergency brake had been installed. If *Giotto* lost its orientation because of an impact, a built-in safety system would try to regain the original position. This system was to show its worth.

But what could hit the probe, which was flying through empty space? The danger lay in the comet's head, which is the immediate source of cometary gas and dust tails. There must be dust particles there, but no one knew how many; no one knew their size. The only certain thing was that they would encounter the probe at very high velocities, because of *Giotto's* orbit. In order to take advance of the earth's orbital velocity, the probe had to be launched in the same direction in which the earth moves around the sun. But, whereas all the planets follow interplanetary traffic rules and orbit the sun in the same direction, Halley's Comet is like someone driving against the traffic. It orbits the sun in the opposite direction, and thus the comet and the probe would be moving in opposite directions. This had two consequences: First, the camera had to form a sharp image of the object as it was speeding past. The Danish astronomer Richard West—himself a discoverer of a comet, which is named after him—compared the task that Keller and his team had to solve with building a camera that had to obtain a sharp picture of the face of the pilot of a Concorde jet as it flew past at a distance of only 30 meters and at supersonic speed. The high velocity of the flyby caused another problem, however. Particles in the head of the comet would impact on the probe at velocities of 68 kilometers per second. The bullet from a rifle travels at a speed of only 800 meters per second. A particle with a mass of only half a gram would pass completely through the probe, should it encounter one. But the probe would also be set rolling, which would cause the aerial to lose the earth from its sights.

But on the evening of 13 March, no one knew what the conditions were like inside the head of Halley's Comet. Around 20:00 Universal Time, *Giotto* passed through the comet's "bow shock," the layer where the density of the solar wind undergoes a sudden jump. In it, the magnetic fields carried by the solar wind are greatly strengthened, and the particles arriving from the sun are strongly heated.

At about 21:03 the color camera began to work. The probe was still 750,000 kilometers away. Only the dust cloud, illuminated by sunlight, could be seen; the nucleus itself could not be recognized. At a distance of 250,000 kilometers from the nucleus, *Giotto* was hit by the first dust grain, but the particle was too small to affect the probe's orientation. The number of impacts was to rise to 120 per second before the dust would have any effect on the probe. The instruments on board established that most of the particles had diameters of a few ten-thousandths of a millimeter. They consisted of so-called organic molecules, that is, carbon and hydrogen compounds. According to other measurements, the comet was losing up to 30 tons of water vapor every second. Metallic elements such as sodium and iron were also detected in the cometary gases.

Then pictures of the comet's nucleus arrived. They showed that it was elongated "like a giant peanut." With a length of 15 kilometers, it was longer than had previously been expected. Its surface was completely black.

There was an air of excitement in the spaceflight center in Darmstadt, West Germany, where signals were received from *Giotto* via Australia, and from which commands were passed to the probe. It took eight minutes for a signal to reach the spacecraft and as long for a reply.

On innumerable television screens, the elongated cometary nucleus was visible at top left—the camera having been programmed to track the brightest part of the image. But that was a bright jet, which lay to the right and below, pointing in the direction of the sun.

So this was the nucleus of the comet (Figure 56) whose gas and dust had been seen by the Chinese in 240 B.C., and whose subsequent return in 164 B.C. had been recorded on cuneiform tablets! It was the object visible over the Catalaunian Plains, where Attila the Hun was defeated in A.D. 451. It was in the sky in 1066, when the Normans invaded England. Johannes Kepler saw it in 1607, and its next return was observed by Halley and Newton. And now we could see this object, photographed in close-up, on our television screens, deprived of its mystique and yet still not fully understood. Another picture, taken from a distance of 1500 kilometers, showed just a part of the nucleus. The camera had again kept the brightest part in the center of the image. We were looking at a hole in the nucleus, out of which streams a jet of gas.

Now the dust particles were beating a tattoo on the probe. In another 20 seconds, *Giotto* would reach the point of closest approach. As they were receiving the pictures, no one in Darmstadt knew that the probe had already lost its orientation long ago. News of the probe's failure took eight minutes to reach the earth. By the time this was realized, *Giotto* had long since flown past the nucleus at a distance of about 600 kilometers. Shortly before, the barrage of dust- and sand-sized grains had set the probe tumbling. The aerial was no longer pointing toward the earth.

The safety system designed for just such an eventuality took 35 minutes to orientate *Giotto* and to re-establish the radio link. The instruments survived the encounter nearly unscathed and continued to send back data for days. Only the camera no longer functioned properly. Images were received, but there was nothing to be seen on them.

More than 50 television networks were linked up for the direct transmission of images. The pictures were shown in so-called false color. In order to emphasize details, the different brightnesses in an image are shown in different colors. For example, yellow might represent the brighter parts of the image, and green darker. On television, this was only rarely explained, and as a result, millions of viewers have since thought that Halley's Comet was as brightly colored as a parrot.

Apart from the camera, the instruments on board *Giotto* appear to have withstood the encounter with Halley. The camera has still not been sufficiently tested to establish the extent of its damage. *Giotto* continued its flight, with its instrumentation shut down. As long as the probe was distant, transmissions to earth needed a lot of power, and therefore tests were postponed in order to save energy for future missions. On 19 February 1990, Giotto was awakened from its slumber. A radio signal was sent to the probe to start a lengthy procedure in which it has been prepared for its new task. The full extent of the damage that the probe suffered in its encounter with Halley's Comet has been determined. On 2 July 1990, the probe passed by

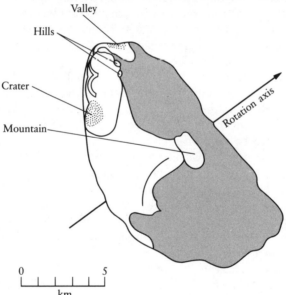

Figure 56. *The nucleus of Halley's Comet (top). The image has been created by superimposing six individual images, obtained by the camera on board Giotto at distances between 3000 and 14,000 kilometers. Sunlight is falling on the nucleus from top left. On the day side, bright jets of dust are erupting from the nucleus. Dust production is restricted to a few centers on the side of the nucleus that is turned toward the sun. Individual features are indicated schematically on the diagram (bottom) (Photo: Max Planck Institute for Aeronomy).*

the earth at a distance of about 20,000 kilometers, which is about one twentieth of the earth–moon distance. The instruments were switched back on, in order to prepare them for a probable encounter with Comet Grigg–Skjellerup on 14 July 1992.

While the equipment on board took its well-earned winter sleep, the teams of researchers on earth were evaluating the data provided by their specific instruments. Initially only general outlines but no individual structures could be seen; modern image-processing, however, has revealed many details. This was achieved by superimposing various individual images obtained at different distances. In Figure 56, the dark, elongated, irregular object is the night side of the cometary nucleus. On the right-hand side we look past the nucleus to dust partially illuminated by the sun. The left-hand edge of the dark area is the boundary between the day and night sides. At the time the picture was taken, it was dawn along that line. To the left of the dawn line we are looking at the bright day side of the nucleus and at dust fountains illuminated by the sun. The diagram in the lower half of Figure 56 indicates the individual features that can be seen in the photograph.

It is not yet clear how the images obtained by the Soviet *Vega* probes agree with those from *Giotto*. The images are difficult to compare, because the *Vega* probes saw more of the nucleus from the day side, rather like the way we see the moon at full moon, whereas *Giotto* saw it like the half moon.

It will take a long time for us to fully understand all the details on the images obtained. In fact, it has only recently become possible to detect details that have been present for 70-odd years in old photographs of Halley's Comet taken in 1910, by re-evaluating the photos with modern image-processing techniques. These procedures enable us, for example, to show that the nucleus rotates around its axis once every 50 hours. The *Giotto* images undoubtedly contain information that will only be recognized at some time in the future — perhaps even when the comet next returns. Therefore, all the information gained in the Halley Campaign in 1986 will be stored on compact discs, so that all the details they contain, even those of which we have no suspicion, will be available for future generations.

Undoubtedly, in 2061, when Halley's comet returns, the old pictures taken by Keller's camera will again be seen on television — but, we hope, not in false color.

▪ *The Cometary Cloud*

Because long-period comets are only close to the sun for a very small fraction of their orbital period, space outside the solar system must be full of inactive comets, with one only occasionally coming in toward the sun. Out there, they are moving with velocities of only a few centimeters per second. The Dutch astronomer Jan Hendrik Oort has estimated the number of these inactive comets as about 100 billion.

Where do all the comets in this cloud come from? Where does their material come from? Where were the gasses and molecules formed that are now frozen in their nuclei? It would seem that comets developed at the same time as the planets. It is possible that the material that then surrounded the sun, and from which the planets were formed, has remained in its original state within cometary nuclei.

The ice in Siberia frequently yields the body of a mammoth, frozen long ago, which provides us with essentially unaltered information about the state of the animal before its species became extinct. The material frozen in comets acts in a similar way. We can probably learn from it what the material that formed the planets was once like.

Comets would not have formed out in the cometary cloud. Oort thinks that they arose much closer to the sun, roughly where the outer planets are found today. Small bodies first formed from the basic material, and these then went to form the planets (see Chapter 12). But the planets perturbed the orbits of the remaining smaller bodies. Even if the latter originally moved in circular orbits, planetary perturbations would cause these to become more and more elongated ellipses, the outer portions of which would eventually lie far beyond the solar system. Out there, however, the orbits are subject to the influence of other stars that may cause them to become circular again, so that the comets no longer visit the inner part of the solar system. In these circular orbits, they form the Oort cloud, which consists of small bodies dating from the time when the planets were formed. Banished to the outermost regions, these inactive comets have slowly followed their gigantic, circular orbits around the distant sun for a vast period of time.

From time to time, another star passes these outer regions, and its gravity perturbs the bodies in the Oort cloud. Some are flung even farther out into space so that they will never return. But others are diverted inward, in the direction of the sun. There, the planets, particularly Jupiter, force many of them into less elongated orbits so that, like Halley's Comet, many return at relatively short intervals. These are the periodic comets whose returns we see regularly.

But we also observe comets that are apparently approaching the sun for the first time. They can be recognized by their extremely elongated orbits, which indicate that they have come from far beyond the outer planets. They also develop particularly bright gas and dust tails. Comets with orbits that do not stretch much farther than the orbits of Neptune and Pluto have frequently approached the sun. As a result they are no longer very spectacular, because they seem to have exhausted their supply of volatile material.

What would happen if one of the cometary bodies that enter the inner solar system were to hit the earth? We know of impact craters that indicate the occurrence of such events (or similar ones) in the past. The Tunguska event of 1908 (see Chapter 3), however, does not appear to have been caused by a comet. From the amount of energy released in the impact, the size of the body has been estimated as around 100 meters — too small to be a comet. It was probably a fragment of a comet.

▪ *Dinosaurs and Comets*

In recent years, it has been suggested that the major impact craters on earth did not occur sporadically, but rather periodically in large "showers." It has been estimated that the earth has been bombarded by particularly large numbers of objects from space at intervals of 20 to 30 million years. This idea has not gone unchallenged, but people have tried to explain other inconsistencies in the earth's history in this way. It would seem that in the past, the extinction of various types of animals occurred in episodes. If large bodies from space hit the earth, clouds of dust produced by these catastrophic events would darken the skies for a long time. Plants would die off, and the dinosaurs would starve.

When impacts of large bodies from space are considered, the obvious suggestion is that comets are responsible. But why would particularly large numbers of comets hit the earth at intervals of 20 to 30 million years? That could only occur if comets became particularly plentiful in the solar system at such intervals. Even though the earth is very unlikely to encounter any of them, over the course of millions of years a few might hit the earth.

When we remember that comets originate in the Oort cloud, then the Oort cloud would have to send comets into the inner solar system at regular intervals. If passing stars initiate showers of comets, then they would also have to pass close to our solar system at regular intervals, thereby affecting the orbits of objects in the Oort cloud and causing them to fall inward toward the sun. But why should stars pass the Oort cloud at regular intervals?

One possibility is that another star is gravitationally bound to the sun and orbits it in a very elongated ellipse. We know that space contains many double stars, where two bodies move in elliptical orbits around their common center of gravity. It has been estimated that a star with about ten times the mass of Jupiter, with an elliptical orbit around the common center of gravity (which in practice lies close to the sun), would cause perturbations in the cometary cloud with the right rhythm and the right strength. Its closest point to the sun would be at about ten thousand times the earth–sun distance, and its farthest point about 18 times as great. That would produce an orbital period of 27 million years. At such a distance, this companion to the sun would pass through the cometary cloud and would send inactive comets into the inner solar system, where they would become active.

Attempts have been made to search for this hypothetical companion to the sun. It would be an insignificant, tiny star, that would only be apparent by its motion. So far, it has not been found. But it has already been given a name: *Nemesis*. The name of anyone discovering a stellar companion to the sun would become immortal. But quite possibly no one will ever gain such fame, because to the Greeks, the goddess Nemesis was the enemy of any good fortune. Many astrophysicists are very skeptical of the theory that the sun has a companion such as Nemesis, of regular showers of comets, and of any effects on the earth's fauna and flora. Above all, it appears very unlikely that any such stellar companion of the sun would orbit the sun for such a long time without being perturbed and torn away from the sun's influence by some passing star.

Whether comets hitting the earth are responsible for the extinction of the dinosaurs or not, innumerable small bodies orbit in the space between the planets. The fact that many of them have at some time encountered larger bodies can be seen from the pock-marked surfaces of Mercury and of many planetary satellites. There are also tiny dust particles existing between the planets. All these smaller objects in the solar system are discussed in the next chapter.

EIGHT

Asteroids, Meteors, and Interplanetary Dust

> The night of the 11th to the 12th of November was cool and exceptionally beautiful. Toward morning, from half-past two, remarkable fiery meteors were seen toward the east. . . . For four hours, thousands of fireballs and shooting stars cascaded down.
>
> ALEXANDER VON HUMBOLDT
> *Account of his journey to Venezuela in 1729.*

■ *A Plethora of Planets*

It was New Year, 1801. In a few hours the sun would rise on the first day of the new century.[1] That night, Giuseppi Piazzi (1746–1826), the new director of the Palermo Observatory, as working at the telescope. While most Sicilians were celebrating New Year, he went on compiling his catalog of fixed stars, which was to contain over 7500 accurate positions of stars. To do this, he measured their positions relative to one another. That night he was working on a field in the constellation of Taurus, where he discovered a star that was not shown on his star chart. Was it an error in the chart? The next night, he saw the object again, but not in exactly the same place. In the interim, it had moved slightly toward the west.

Piazzi followed the puzzling star for another six weeks. Every day, the object moved slightly farther west. But then its movement became slower and slower. Finally, it started to move eastward. Before news of Piazzi's remarkable observations had reached the rest of Europe, Taurus was no longer accessible to Piazzi's instrument, which could only observe stars as they crossed the north–south line. Whatever the object was, it was soon in the daytime sky and could no longer be seen. In the German city of Braunschweig, the 23-year-old Carl Friedrich Gauss (1777–1855) used Piazzi's observations and devoted his attention to the remarkable star.

By the beginning of the nineteenth century, it had been clear for a long time that all celestial bodies had to move in accordance with Newtonian mechanics. What sort of orbit would a body be following in order to agree with the positions where Piazzi had seen it at different times? Gauss calculated that Piazzi's star was actually moving in an elliptical orbit. Even more, he was able to predict where Piazzi's star could be recovered. Toward the end of 1801, people were hoping to find the mysterious object at the position predicted by Gauss. Friedrich Wilhelm Olbers (1758–1840)[2] nearly became the first to find it again when he discovered Piazzi's object on 1 January 1802, at the spot predicted by Gauss; however, he had been beaten by one day by the director of the observatory on the Seeberge near Gotha, Franz Xaver von Zach (1754–1848). It was a time when directors of observatories still went up into the domes and actually looked through the telescopes themselves.

The new planet was named *Ceres*. While he was following it in the sky, Olbers found a second faint planet on 28 March 1802. It was named *Pallas*. Then more were discovered. At the Lilienthal Observatory close to Bremen, the theologian Karl Ludwig Harding (1765–1834) found *Juno;* and then it was Olbers's turn again when he discovered *Vesta*. All these new, faint planets orbited the sun between Mars and Jupiter. The next to play a part was the postmaster Karl Hencke (1793–1866) from Pomerania, who continued the hunt for planets. He searched the skies in vain for fifteen years before being successful. For finding two new planets, he was granted a pension by the king of Prussia and a gold medal by the king of Denmark. The list of amateurs was increased by a Parisian painter. Hermann Goldschmidt (1802–1866) was a melancholy person, who, when he was in the depths of depression, tried to find something else to do in order to get new ideas. By chance, he attended a lecture by Le Verrier (see Chapter 5) at the Sorbonne. As a result, he began searching the sky with his small telescope. Goldschmidt was over 50 when he found his first planet, but he discovered a total of 14. He was awarded many honors, prizes, medals, and finally a pension. Following his death, a lunar crater was named after him.

In those days, it was necessary to compare the view through a telescope with a star chart in order to discover a new planet. But then photography arrived. It became possible to compare plates of the same area of sky taken at different times with one another, and to detect any changes directly. A point of light that had changed its position in the meantime was immediately obvious. Now the number of newly discovered planets rose dramatically.

Today, almost exactly 3000 are sufficiently well-observed for their orbits to be known. These asteroids have been tidily provided with names and are listed in catalogs. They can always be found, because their positions on the sky can be calculated for any particular point of time. There is a far larger number that have been observed but not followed across the sky for a long enough time to determine their orbits, there are doubtless numerous such objects in the solar system.

These small bodies discovered since Piazzi's time are called *asteroids,* because they are so small that even through a telescope they appear starlike, or *planetoids,* because they behave like proper planets, but are much smaller. Sometimes they are simply called *minor planets.* They are, in fact, very small. Mercury has a diameter of about 5000 kilometers, but even the largest minor planet is no more than a few hundred kilometers across. We undoubtedly only see the largest, and there is probably no lower limit to their size. Most of them orbit the sun in a region whose diameter is about three times that of the earth's orbit. The asteroid belt, as this region of minor planets is known, lies outside the orbit of Mars and inside that of Jupiter.

In a certain sense, they are unlike the "true" planets. It is not just that they are considerably smaller, but also that their orbits are far less ordered. The orbits of the major planets are close to circles, but among the minor planets some have orbits that depart greatly from a circle. At the point on its orbit nearest to the sun, Icarus, for example, is far inside the orbit of Mercury, but at its farthest point it is outside the orbit of Mars. In Chapter 3, we mentioned the minor planets that come particularly close to the earth: Eros and Amor. They were used in the past to determine the distance between the sun and the earth.

The asteroids do not appear to be spherical bodies. Their shape probably resembles that of the Martian moon Phobos, because their brightness changes rhythmically. They become brighter and fainter over periods of hours. If, like all the bodies in the solar system, they rotate, and if they are irregular in shape, then we would sometimes see more and sometimes less of the surface illuminated by the sun. Even with our largest telescopes, they appear just as points of light, and their brightness varies in accordance with their period of rotation. For example, Ceres, Piazzi's discovery that New Year's Day, appears to rotate once in 9 hours and 5 minutes. Icarus takes just 2 hours and 16 minutes to complete a rotation.

Minor planets exhibit another irregularity. All the orbits of the major planets lie in approximately one plane. The individual orbital planes have low inclinations to one another. The asteroids are quite different: Their orbits frequently lie at large angles to the earth's orbital plane (the ecliptic). For example, the orbit of Hidalgo is inclined at about 45° to the plane of the ecliptic. Orbital inclination is the reason that Icarus cannot collide with the earth. Although it sometimes comes within the earth's orbit, it never actually intersects it.

Not all the minor planets have unusual eccentricities or highly inclined orbits relative to those of the major planets; some are even extremely well behaved. In Chapter 2, we talked about the Trojans. These asteroids orbit the sun in almost perfect circles, such that Jupiter, the sun, and the Trojans always form the corners of an equilateral triangle. This is not strictly correct. Each Trojan orbits in the vicinity of a Lagrangian point, which forms an equilateral triangle with the sun and Jupiter. Although many other asteroids follow almost random orbits, the Trojans are strictly linked to the motion of Jupiter.

We know a lot about the orbits of asteroids, but we do not know very much about the bodies themselves. It is not surprising that they are not spherical. We have already discussed in Chapter 4 the greatest height that mountains on any planet can attain, and we calculated how high a pile of rocks can be before it begins to sink under the influence of gravity, which also tries to force planets to become spherical. There is practically no gravity on an asteroid. On a body 100 kilometers in diameter, a 65-kilogram man would weigh the equivalent of half a kilogram on earth! For just the same reasons that a wardrobe, say, taken up into space would not turn into a sphere, asteroids also remain as angular as they were originally.

How did the minor planets arise, and where do they come from? Their orbital peculiarities led Olbers to suggest that at some time in the past there was a large planet between Mars and Jupiter that exploded, or was somehow broken up, and that the asteroids are its remnants. Nowadays, however, it is believed that they were formed at the same time as their larger neighbors. We

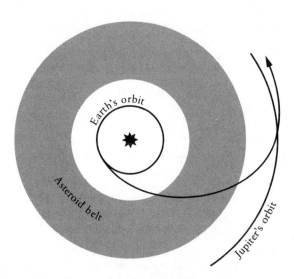

Figure 57. *Thousands of asteroids orbit in a wide belt between Mars and Jupiter. When* Pioneer 10 *traveled to Jupiter, it had to cross this dangerous region. Since then, many probes have crossed the asteroid belt without any collisions.*

shall probably know more about their origin when we are able to see one in a close-up image. We may perhaps soon have the opportunity to do so.

Space probes that have reached the outer planets have passed the region where asteroids are found. The probe *Pioneer 10* (see p. 182) was the first to cross the asteroid belt in 1971–1972 (Figure 57). At the beginning of 1985, NASA announced plans to launch the Jupiter probe *Galileo*. The idea is to divert the probe slightly so that, in crossing the asteroid belt, it comes within 10,000 kilometers of the minor planet Amphitrite. This should allow the surface to be mapped, and the mass, density, and rotation of the planetoid to be determined. It is estimated that the diversion will cost 20 to 25 million dollars. Following the *Challenger* accident in January 1986, the plan had to be dropped, because the launch of *Galileo* would be delayed, leaving no time for the diversion.

But we already know that we do not always have to go into space to study extraterrestrial material. Some material comes to us from space.

▪ *Shooting Stars*

As August begins, they turn up individually. Then more and more of them arrive every night. The peak comes on 12 August, or a day earlier in leap years. On that night, shooting stars, which are known by the romantic name "tears of Saint Laurence", flit across the sky from a point between the constellations of Cassiopeia and Perseus.

But shooting stars, properly called *meteors,* do not just arrive on 12 August. In much smaller numbers, they can be observed any night. Many occur in showers, and many more individually. The general public often confuses them with comets. But whereas a comet rises and sets with the stars and moves across the background of the fixed stars in days or weeks, meteors dart across the sky. Anyone who is not looking up at the correct moment misses them. Meteors and comets are completely different objects. And yet they do have some things in common, which we will discuss later.

Frequently, a meteor is a single, random event. But most meteors come in showers. The members of such a shower all appear to come from a single point of the sky, which is known as the shower's *radiant*. Although they do not arrive all at once, but instead come separated by intervals that are typically a few minutes, they are rather like an exploding firework, shooting sparks out in all directions (Figure 58).

Astronomers call the showers after the Latin name of the constellation from which they appear to come. Those seen on 12 August have a radiant in Perseus and so they are called *Perseids*. There are also the *Lyrids* (from Lyra), the *Leonids* (Leo), the *Piscids* (Pisces), and others. Alexander von Hum-

Figure 58. *Shower meteors all appear to come from a single point on the sky, known as the radiant. The reason for this is the effect of perspective, which makes parallel lines appear to converge to a point.*

boldt's account quoted at the beginning of this chapter concerned the Leonids. All meteor showers arrive at specific times of the year: the Lyrids around 22 April, the Piscids on 12 September, the Leonids in mid-November.

What are shooting stars and fireballs? Like Tycho Brahe, who in 1577 tried hard to measure the distance of a comet from the earth, two students from Göttingen tried in 1798 to find the distance of meteors above the surface of the earth. Since a shooting star does not actually occur very high in the atmosphere, when seen from two points its path appears in different places in the sky. From this difference, the height can be determined. In order to do this, the two students, Johann Friedrich Benzenberg (whom we already met in Chapter 2) and Heinrich Wilhelm Brandes (1777–1834), observed meteors from two sites near Göttingen, plotting the observed paths of the meteors on star charts and recording the Universal Time, so that they would be able to determine which of their observations applied to the same meteors. They were successful and found that meteors occurred at heights of between 35 to 125 kilometers. Another result of their work was that they obtained the velocities of the meteors, which amounted to an impressive 30 to 45 kilometers per second. In addition, they confirmed that all shooting stars were moving downward, toward the surface of the earth. So people were dealing with objects arriving from space and coming down toward the earth.

The results obtained by the two Göttingen students were so important that Alexander von Humboldt mentioned them in the third volume of his work *Kosmos,* which appeared in 1850. Both students continued with science. We already mentioned Benzenberg's work on dropping lead spheres from the top of St. Michael's Church in Hamburg. He later taught mathematics in Düsseldorf and worked in the survey office. He bequeathed his private observatory to the town of Düsseldorf. Brandes eventually became a professor of mathematics and later a professor of physics.

Why do meteors occur preferentially in showers, and why do the shooting stars in a particular shower appear to come from the same point on the sky? The answer to the last question is quite easy. We know that many things appear different to the eye than they actually are. Railway lines appear to come together at a single point on the horizon because of the effects of perspective. When we look along parallel lines, they appear to get closer together the farther away they are. In reality, they remain the same distance apart. The same applies to the tracks of meteors in the sky. The observer gets the impression that all are coming from a single point in the sky, the radiant. In fact, all the meteors in a shower are traveling along parallel paths when they encounter the earth's atmosphere. When we look at the radiant, we are looking along their paths, just as if we were standing on a railway track and were looking along the rails.

But why do individual showers arrive on the same dates in the year? The bodies that vaporize in the atmosphere are not evenly spread throughout space. There are "highways" along which meteors travel in parallel paths. When the earth crosses one of these highways as it travels around the sun, it encounters a large number of these bodies, and we see a meteor shower. Because the earth returns to the same point on its orbit a year later, it passes through the stream on the same date, and the shower is repeated.

I have just spoken of highways along which meteors are moving in parallel paths. How does this traffic pattern come about? Two astronomers discovered the answer almost simultaneously in the last century. One of them was Giovanni Schiaparelli in Milan, whom we met in connection with the Martian canals. He tried to determine the orbits of meteors in space. All the meteors' orbits strongly resembled parabolas or, at the very least, extremely elongated ellipses. Comets have precisely these types of orbits. Schiaparelli then discovered that a comet that appeared in 1862 had nearly the same orbit as the Perseid meteors. Exactly the same result was obtained by the American mathematician Hubert Anson Newton (1830–1896), who taught at Yale University. Other astronomers soon found agreement between the orbits of certain known comets and the annually recurring meteor showers. For instance, the Leonids follow the orbit of a comet observed in 1866.

Whenever the earth passes close to the orbit of a comet, groups of glowing particles streak through its atmosphere, and we observe a meteor shower. Even when the comet is at a completely different position in its orbit, we can still detect the meteor shower (see Figure 59). Comets pollute both their immediate neighborhood in space with debris, and also their orbits. As a result, meteor particles that a comet has released are to be found along the entire length of the elongated elliptical orbit, as well as in the immediate vicinity of the comet. The rubbish left behind by comets remains invisible, until the earth happens to pass through the stream, when the particles begin to glow as they strike the atmosphere.

We know nothing of many meteor streams that are moving along cometary orbits around the sun, because their paths are well away from the earth and therefore they do not encounter the earth's atmosphere. But we do come close to the orbit of Halley's Comet twice a year, when we encounter meteor particles from the comet. In May, we encounter those moving away from the sun. The meteors have a radiant lying in Aquarius and are therefore called *May Aquarids*. In October, we encounter those that are approaching the sun from the far reaches of the solar system, the *Orionids,* which appear to come from a point in the constellation of Orion.

Not all meteors belong to showers. There are also stray individuals. On a moonless night, it is usually possible to see between four and six of them every hour. Because of the very nature of these luminous objects, they can generally only be seen in the night sky. Only in very rare cases are any so bright that they can be seen in the daytime sky. Thus, many regular showers went undiscovered. The meteors in a particular shower only occur at a specific time each year, but when the constellation from which they arrive lies in the daytime sky, we are quite unable to observe the event. Recently, a method of observation has been found that enables us to detect meteors in the daytime. We shall talk about this shortly.

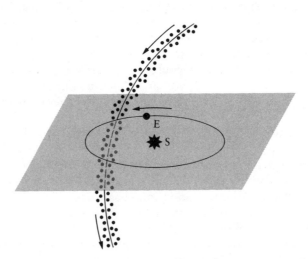

Figure 59. *Many meteors follow cometary orbits in space. If the earth (E) comes close to such a track while it orbits the sun (S), it encounters a swarm of particles all following parallel paths, and a meteor shower is observed. Since the earth always approaches any individual cometary orbit at the same time each year, the meteor shower is always observed on the same nights of the year.*

▪ *Glowing Rocks*

What happens when a body encounters the earth's atmosphere at a speed of, say, 30 kilometers per second? We know that astronauts have to slow down their spacecraft greatly when they return to earth, so that it will not burn up as it enters the shell of air surrounding our planet. A heat shield is required to counteract most of the heat that is produced. Likewise, an unprotected meteor is rushing to meet its doom. The body approaches with supersonic speed, and the air—even though it is very thin in the outermost layers of the atmosphere—is quite unable to move aside. It is carried along by the body and becomes greatly compressed, which causes the air in front of the body to be strongly heated. This heat is transferred to the meteor, which begins to glow, to shine, to melt, and eventually to vaporize. Along its path, it leaves vaporized material. Frequently, the glow from the material left behind a meteor can be seen in the sky for several seconds afterwards. Yet the particles that result in meteors are very tiny. It is estimated that the mass of a meteor of average brightness is less than one gram. In nearly every case, the whole of the particle is vaporized in the upper layers of our atmosphere.

The temperatures attained by the air around a meteor reach several thousand degrees Celsius. Such a heat is not only dangerous to the incoming particle, but also affects the atoms in the air itself. The heat splits electrons away from their parent atoms. Whereas the atoms and molecules in the air were previously electrically neutral, positively charged particles remain along the meteor's path. These are ions, or atoms from which one or more electrons have been removed. When the latter recombine with the atoms, light is emitted. As a result, it is not just the meteor particle that glows but also the air surrounding it.

Free electrons are found on earth in the metals. There, not all the electrons are bound to atoms, some remain free between the atoms. This is why metals are good conductors and thus reflect radar waves particularly well. For the same reasons, meteors are good reflectors of radar waves, and they do this equally well whether it is day or night. Since the ions and electrons require a certain time to recombine, the whole track behind a meteor is reflective, not just the point where the particle is at that particular instant. Using radar, we can detect meteors by day or by night.

One only has to turn the aerial toward the sky and to send out radar waves. Wherever a particle from space has entered the earth's atmosphere the radar waves will be reflected back to their point of origin. The distance to the event can be determined from the time taken for the echo to return and from that its height in the atmosphere. The path of the meteor in the earth's atmosphere can also be determined by simultaneous observation by more than one radar station. This is how meteor showers have been discovered that would have remained completely unknown with only visual methods.

We now know more or less what happens when a body encounters the earth's atmosphere at cosmic velocities. But what are these particles that are flying around through space, either as stray individuals, or in streams clustered along the paths of comets?

▪ *Stones from Space*

There are old accounts of people being killed by meteors, but most of these deaths seem to have been caused by lightning. On 30 November 1954, a 3.8-kilogram body crashed on the house of Mrs. Hodges in Sylacauga, Alabama. The body smashed through the roof, hit a roof truss, broke through the ceiling, hit a radio, and injured Mrs. Hodges, who was lying in bed. The body had been considerably slowed down by the collisions. Thanks to this fact and two quilts, the lady required hospital treatment for only minor injuries.

Certainly not all cosmic bodies are vaporized when they encounter the atmosphere. There is often a remnant that reaches the earth's surface. We already discussed in Chapter 3 some of the gigantic events that have occurred, such as the Tunguska event in Siberia and the one that caused the huge crater where the German town of Nördlingen is located today. But apart from the few rare major events, many smaller rocks reach the surface. To differentiate these from the luminous phenomena in the sky, the bodies that reach the surface are known as *meteorites*. Many meteorite falls are known, but meteorites often fall unseen and lie among ordinary terrestrial rocks, only being recognized by experts. Many meteorites do attract the attention of laymen because they consist of iron. Iron meteorites are much rarer than stony meteorites.

It has been known for many years that "stones" fall from the sky. On 7 November 1492, a stone fell, accompanied by loud thunder, in a field in Alsace. The meteorite researcher Ernst Chladni (1756–1827) described the stone, which was later preserved in a local church, as "dark gray, solid, and with a denser core than many others. It contains none of the dark grains found in most [meteorites], but instead pure iron, iron sulfide (some of which occurs in veins), and other fractions that are sometimes a whitish-grey, and sometimes resemble olivine. Among the iron sulfide, a reddish-gray layer appears to indicate the presence of nickel."

Recently, a Canadian research team reported on the results of nine years of monitoring the sky with 60 cameras. If the tracks of meteors against the sky are recorded from several different sites on the ground, the paths of the objects in the atmosphere can be determined, together with any possible impact points. This allows a search of a specific area. Their haul, translated into a figure for the whole of the earth's continents, showed that about 5800 meteorites larger than 100 grams fall on land every year. The authors estimated that over the whole earth, one roof will be damaged by a meteorite every year, and that on average, one of the earth's five billion inhabitants will be hit by one every nine years. But the fall mentioned earlier is the only documented case.

Thousands of meteorites have been found. They always appear to be the remnants of sporadic meteors, and none are known to have come from a meteor shower. Modern methods of analysis allow the age of meteorites to be determined from the amount of uranium and thorium present in the rock. To be more accurate, it is the time since the body became solid that can be determined. The ages found are about 4.6 billion years, the same age as the sun and the earth. Meteoritic material appears to have solidified at the time our solar system originated.

It is also possible to determine how long the bodies have been moving through space as separate bodies, before encountering the earth. This is possible because intense particle radiation pervades the whole galaxy, and therefore also interplanetary space. By this, I mean the so-called cosmic rays. Every grain of dust orbiting in space is hit from time to time by cosmic-ray particles. When the grain is irradiated, some of its atoms are split. The products of the fission can be detected chemically. It is therefore possible to use the products in meteorites to find out how long the bodies were exposed to cosmic rays. The result is surprising. Most of them have been bathed in cosmic rays for only about one-tenth of their lives. For nine-tenths of the time, they must have been protected from the radiation. What, however, protected meteorites from cosmic rays for most of their lives? It would take a barrier of 1 to 2 kilometers of meteoritic-type material to shield the bodies from cosmic-ray bombardment. Did meteorites spend most of their time deep inside a comet or a minor planet, having been released from their parent body only a "short" time ago? Are meteorites extinct comets or remnants of asteroids? We still have no definite answers to these questions.

The space between the planets does not contain only the rocky bodies of the minor planets, the icy objects that are comets, and the many smaller bodies that burn up as meteors in our atmosphere or fall as meteorites. There are also fine particles of cosmic dust, minute grains of which orbit the sun just like true, major planets.

■ *Illuminated Dust*

You have to make an effort to see the faint glow of light that appears after sunset in spring above the western horizon like a slanting triangular patch, several handwidths across against the dark sky, its apex pointing toward the south. A corresponding phenomenon can be observed in the autumn at dawn before sunrise. This is the *zodiacal light*. It is produced by a disk of dust that surrounds the sun and reaches out beyond the earth. The illumination comes from sunlight that is scattered by the dust particles. The diameter of the particles is between one-tenth and one-thousandth of a millimeter. A total of about one-hundred-millionth of the mass of the earth in the form of the finest dust appears to orbit the sun in this disk. When we see the zodiacal light, we are looking sideways through the densest part of the disk.

It is thought that this dust consists of the remnants of comets and probably also of tiny fragments from asteroids. When we recall that the moon, Mercury, and the Martian moons are covered with impact craters, we can expect asteroids to be repeatedly hit by meteorites. The surface of an asteroid is probably as pock-marked as Phobos. Even when a very small meteorite hits a minor planet, most of the fragments will be flung into interplanetary space because of the very low gravity. Thus, we can expect the whole of the solar system, not just cometary orbits, to contain tiny bodies, down to the size of the finest dust.

This has recently been confirmed. Once again, it was a space probe that observed something no one had ever seen before. The satellite *IRAS*, launched in 1983, was a collaborative effort by the United States, the Netherlands, and Great Britain. It "sees" heat radiation,—in other words, it detects infrared wavelengths. Its name is an acronym for Infrared Astronomical Satellite.

In addition to many other discoveries, it detected infrared radiation from dust particles in a broad ring around the asteroid belt. This ring is show schematically in Figure 60. Tiny particles of dust orbit the sun within this ring. Although the ring extends out of the plane of the planetary orbits on both sides, every particle moves in a (plane) elliptical orbit, with the sun at one focus. The orbital inclinations of the innumerable dust particles are such that, taken together, the particles fill the tubular region shown in the diagram.

It is estimated that about 1 trillion tons of material orbits the sun in this ring. Where does all this dust surrounding the asteroid belt come from? If all the dust were to be assembled into a solid body, it would produce an asteroid about 16 kilometers in diameter. At some time in the past, did two minor planets collide in the midst of the asteroid belt, pulverizing themselves and giving rise to the dust that we now see? Or does this dust originate in comets that Jupiter perturbed into orbits close to the sun and whose most distant portions lay in the region of the minor planets? Has *IRAS* observed dust released from these comets?

Cosmic dust also lands on earth. It can be recognized from deep-sea ooze deposits that are recovered and brought to the surface for investigation. In oceanic ooze, a particularly interesting form of aluminum, aluminum 26, is found. It is a radioactive substance and decays into atoms of magnesium in 740,000 years. An atom of aluminum 26 is produced when material is

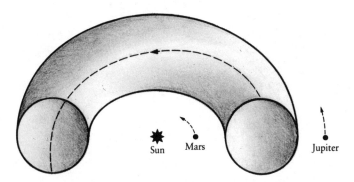

Figure 60. *In the ring of dust encircling the asteroid belt, discovered by* IRAS, *the individual particles move in elliptical orbits around the sun. Since each orbit has a different inclination, they fill the whole of the tubular region. Any given particle that lies at the top of the tube* (right *in the drawing), at one point in its orbit will lie at the bottom on the opposite side of the sun* (left). *The diagram is not true to scale. The "tube" is shown here schematically to make the geometrical layout clearer.*

bombarded by fast cosmic-ray particles, or by the energetic particles produced in solar flares. When the material is protected from energetic radiation by other material for millions of years, aluminum 26 decays. The aluminum-26 enriched material in oceanic ooze must have arrived "recently" from space, because the seas do not allow any particles that could produce aluminum 26 to penetrate to the seafloor. One could say that aluminum 26 still "smells of the sea," but the sea in this case is interplanetary space. From this information, it has been possible to estimate that thousands of tons of material that were recently exposed to cosmic radiation fall on earth each year.

NINE

The Failed Sun

A red spot shows clearly that Jupiter rotates.

JOHANNES KEPLER, 9 January 1611

The planet Jupiter has only about one one-thousandth of the mass of the sun, but it is still 318 times the mass of the earth. It is the largest of the planets; it forces comets to adopt new orbits and holds the Trojan asteroids around two of the Lagrangian points that it forms with the sun (see Chapter 2), causing them to orbit the sun with the same period as its own orbit, namely 4332.6 earth-days or 11.86 earth-years.[1] Its orbital radius is so large that it takes sunlight 43 minutes to reach it. Even under the most favorable conditions, radio signals from Earth take 35 minutes to reach space probes in the vicinity of Jupiter, and any responses take just as long to return.

Because of its long orbital period, Jupiter only moves a short distance along its orbit during the course of a terrestrial year. Thus, oppositions occur at average intervals of one year and one month; to be precise, Jupiter is at opposition every 398.9 days. Since neither Jupiter nor the earth travel in perfectly circular orbits, the distance is not the same at every opposition. A particularly favorable opposition occurred on 18 October 1987, and the next one will be on 23 October 1999. In both cases, the distance between the earth and Jupiter is only 591.5 million kilometers.

In the middle of the seventeenth century, Giovanni Domenico Cassini (see Chapter 4) made detailed observations of Jupiter through his telescope. He noted that belts were visible on the disk and that these changed with time. He also saw that the planet was rotating, because a spot—which he was actually able to follow from 1665 until 1691—regularly disappeared behind the disk of Jupiter and reappeared on the other side after a certain length of time. From this, Cassini deduced a rotation period of 9 hours and about 55 minutes. He also noted that the disk of Jupiter was not circular, but elliptical. In fact, the smaller diameter is about 7 percent less than the larger. It was not long before it was decided that the flattening of Jupiter's disk was a result of the planet's rotation. Jupiter's appearance changes rapidly, spotlike markings arise along the parallel belts, are carried along by the rotation, and then vanish again. Jupiter affords a very interesting program for observers and remains a favorite object of study for amateur astronomers.

Herr Meyer and Jupiter

As Herr Meyer looked out of the window of the base, instead of the sky, he saw a delicately colored, pinkish surface rising above the horizon. Variously colored darker belts, separated by brighter zones, could be seen lying at an angle to the dark line of the horizon that separated Gallia from the sky.

The night appeared to be significantly darker than it had been when they observed Mars. With increasing distance from the sun, less material was being released from the surface of the comet's nucleus by sunlight. The dust fountain on the horizon could hardly be seen except when the sun was high in the sky.

"Gallia will pass through the middle of Jupiter's system of satellites at a height of 50,000 kilometers above the highest cloud tops," said Professor Rosette. "The planet's gravity will change Gallia's orbit, but I have already calculated all the effects. We shall escape from Jupiter's gravitational field again, and Gallia will not be added to its harem. While we are in the vicinity, we should make the most of it." Professor Rosette picked up a pair of binoculars and looked out of the window. Herr Meyer did the same.

So that was Jupiter. Among the varied play of colors on the planet, Herr Meyer could see regular features. On one side of the equator, clouds were arranged like a string of pearls, reminding Herr Meyer of terrestrial cumulus clouds. On both sides of the equator, the whole disk was sprinkled with small whitish spots (Figure 61). Within their interiors, they showed various light tints. Herr Meyer could make out their spiral structure, looking almost as if watercolors had been mixed together inside them.

"The spots are whirlwinds," said the professor. "The prime example is the Great Red Spot."

This oval reddish spot, Jupiter's trademark (Color Plate 4), had already attracted Herr Meyer's attention. It lay halfway between the

Figure 61. *Jupiter photographed by* Voyager 1 *from a distance of 40 million kilometers. The cloud belts, known from observations from Earth, can be seen and allow the planet's rotation to be determined. The Great Red Spot is visible in the southern (lower) half of the picture (see also Color Plate 4). Ganymede, one of Jupiter's satellites is at the bottom left of the picture (Photo: NASA/JPL).*

equator and one of the poles, embedded in the system of parallel cloud belts, between a broad white zone and a thin dark belt of clouds.

"For a long time it was thought to be a solid ice floe, floating on the liquid surface of Jupiter," the professor continued. "Even from Earth it is possible to see how the clouds flow around it. The spot is a gigantic whirlwind, a tornado, which rages in one place, without dying away. When terrestrial cyclones cross onto land from the sea, they slow down and die out. But on Jupiter, there are no continents, and the Great Red Spot is a very persistent hurricane. The whole of the spot rotates with the rest of the planet in one Jupiter-day, which is about ten earth-hours. The gas currents inside that whirlwind take about six earth-days to circulate around once inside the spot. The diameter of the hurricane is about 40,000 kilometers; so the earth would fit inside the Great Red Spot with plenty of room to spare."

Through his binoculars, Herr Meyer could see even finer structures. White bands, looking like streams of milk within a reddish fluid, appeared to flow past the Great Red Spot, which itself showed faint flow lines. The white current formed regularly spaced eddies. The motion inside the Great Red Spot itself appeared to be anticlockwise.

Then, looking at the dark sky, Herr Meyer discovered a tiny illuminated spot, which was obviously moving against the background of stars.

"Have you found Amalthea?" asked Professor Rosette. "It was the innermost of Jupiter's moons until the *Voyager* probes found two more that orbit even closer to the planet. Amalthea itself is an irregular lump of rock, again shaped rather like a potato, like the Martian moon Phobos. Its longest diameter is about 260 kilometers. Amalthea was discovered in September 1892. Tidal friction has affected its rotation so that it now always turns the same face toward Jupiter. It revolves around the planet in nearly 12 hours and rotates around its axis in the same period of time. Its long axis thus permanently points toward the center of Jupiter. But, if I have not made a mistakes in my calculations, Io should be rising about now."

Shortly afterwards, an orange-colored disk, about as large as the full moon appears to us, rose over Gallia's horizon. With his binoculars Herr Meyer could see white and yellowish-red markings and, between them, black points, looking like pores on the otherwise smooth skin of Jupiter's satellite.

"What we are looking at is a crust of sulfur, which is covered in places by sulfur dioxide snow." It was a strange, dead world that presented itself to Herr Meyer's view (Color Plate 5). Suddenly, a jet of material shot up from one of the pores close to the edge of the disk. It spread out, and whatever it was that had been shot out of Io slowly fell back to the surface. The fountain of material, seen from the side, looked rather like an open umbrella.

"There are many volcanoes on Io, but no impact craters. The material ejected from the volcanoes probably immediately hides any craters that do occur. You can see traces of earlier volcanic eruptions through your binoculars. The volcanoes are surrounded by a ring, which has obviously been formed where the ejecta have fallen back onto the surface." Herr Meyer could indeed see a ring around the dark spot.

In the meantime, they had flown so far past Jupiter that Herr Meyer could now see the half-illuminated disk. On Gallia, it was half Jupiter. By now, Io had moved away, and Herr Meyer could no longer see it out of the window. Then he noticed two other disks, which rose almost simultaneously.

"Europa and Ganymede," commented Professor Rosette. "Ganymede is the largest of Jupiter's satellites, and as seen from Earth, Europa is the brightest. It may have a crust of ice some 100 kilometers thick. The surface of this satellite is probably simply a frozen ocean." Through his binoculars Herr Meyer could see that the otherwise exceptionally smooth disk was covered with a network of dark lines, intersecting one another at all angles, just as if giant spiders had spun their webs all over the surface (Figure 62).

"If we are really looking at a crust of ice, then it would seem that after the crust formed, the moon expanded, cracking its icy armor into innumerable fragments. There are very few impact craters on Europa—none with a diameter greater than 50 kilometers. This suggests that the crust is not very old. It may be possible, however, that water wells up repeatedly through the rifts, burying any impact craters beneath everlasting ice."

"Ganymede, on the other hand," continued Professor Rosette, pointing to the other disk visible in the sky, "consists of a relatively dark terrain, on which there are impact craters. Many of the grooves that spread out from the craters have been broken by faults. Ganymede does not appear to be completely quiet geologically."

After a pause, the professor continued: "Now we can see Callisto. It is just rising above the horizon. As it is about 2 million kilometers from Jupiter it will come very close to us, because by now, we have moved farther away from the planet. See if you can recognize the Valhalla region and its gigantic crater."

Herr Meyer saw the feature immediately. It was a spot surrounded by concentric rings like those on a target. He could make out about twenty thin, circular rings, more or less parallel to one another and equally spaced, which encircled a large crater at their center.

Figure 62. *On the morning of 9 July 1979, Voyager 2 photographed Jupiter's satellite Europa. Apart from the network of lines covering its surface, the lack of impact craters is very noticeable. No sign of any ruggedness can be seen along the smooth edge of the moon. It would seem that, when any impact crater is formed, it becomes covered by water, which then freezes. If water is still liquid within Europa, it must be kept fluid by some heating mechanism* (Photo: NASA/JPL).

While Herr Meyer was wondering what sort of impact might have caused the crater and so many rings, the French professor interrupted him.

"Let's take another look at Jupiter. We have now entered its shadow and can see the dark hemisphere, unilluminated by the sun." The planet had become smaller, because Gallia had moved away from it, and Herr Meyer could cover it with his hand held at arm's length. They were in the middle of the shadow, and Gallia was undergoing a solar eclipse caused by Jupiter. Although the surface of the planet was no longer illuminated by the sun, the light falling on Jupiter's day side was refracted by the atmosphere and was bent around onto the dark side. Jupiter appeared to be surrounded by a bright halo. But even the nonilluminated disk was not completely dark; in the polar regions, Jupiter's disk was glowing.

"Aurorae," said the Frenchman. Elsewhere, Herr Meyer saw the sudden flash of lightning—and then he discovered the ring.

He had not recognized it immediately, because Gallia was precisely in the plane of the ring, so that the ring was being seen from the side. Only two luminous straight lines were visible, in line with the equator and stretching out into the dark sky on both sides of the planet. Each was slightly less than one planetary radius in length.

"The outer edge of Jupiter's ring," said Professor Rosette, "is about 55,000 kilometers above the highest cloud tops. Therefore, Gallia must have flown through the ring. The fact that we did not notice anything indicates that the lumps of rock of which it consists must be very thinly scattered in space. Their diameters are estimated to be at most a few meters. The whole ring is probably no more than 30 kilometers thick. Because of the light from Jupiter, it cannot be seen from Earth."

The next morning Herr Meyer realized that his dreams must have been inspired by the *Voyager* pictures of Jupiter and its moons.

▪ *Pioneer and* Voyager

So far, Jupiter has been visited four times by spacecraft from Earth (Figure 63). The visits began with the probe *Pioneer 10*, which was launched on 2 March 1972. The probe traveled for 21 months before it flew past the planet's upper cloud layers on 4 December 1973.

It had not just had a long journey, but also a dangerous one, because the probe had passed through the asteroid belt (see Chapter 8) and had flown dangerously close to the minor planet Nike, but without any collision. Named asteroids are relatively harmless, because we generally know their orbits and when and where they are likely to be; thus, a dangerous encounter can be avoided. What are dangerous, however, are the smaller objects that are invisible to us. Even a tiny rock the size of a pea can be dangerous when it hits a spacecraft with a velocity several times that of a rifle bullet. But *Pioneer 10* survived the trip unscathed and reached Jupiter and its moons. The planet has a strong magnetic field, and the probe's instruments investigated the interaction between the solar wind and Jupiter's magnetic field. Particles are trapped within the latter just as they are in the earth's radiation belts. In December 1973, *Pioneer 10* reached the region containing the planet's satel-

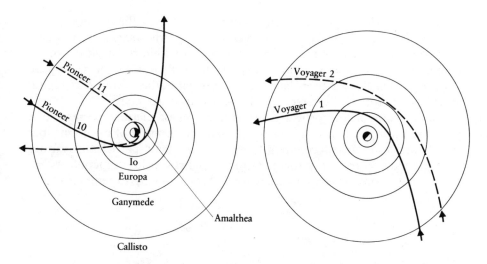

Figure 63. *The probes* Pioneer 10 *and* 11 *and* Voyager 1 *and* 2 *penetrated the region containing the Galilean satellites.* Pioneer 10 *went within 131,400 kilometers of the planet's cloud layers, which corresponds to one sixth of Jupiter's radius.* Pioneer 11 *went even closer, to within 42,000 kilometers. The* Voyager *probes only approached to within some 278,000 and 645,000 kilometers of Jupiter's upper cloud layer.*

lites, passing through the middle of the radiation belts. At that point, the whole mission became jeopardized, though not because of the effects of the intense particle radiation; the danger was on earth. A strike by the operators of the diesel generators at the ground station in Australia nearly resulted in the loss of the data that had been gathered during the six hours of the probe's closest approach to Jupiter.

A year later, *Pioneer 11* passed closer to Jupiter than any of the planet's satellites then known, even closer than the (provisionally) innermost moon, Metis, which was found by a later mission and which orbits 56,000 kilometers above Jupiter's cloud tops. As seen from the probe, Jupiter's disk must have covered half the sky.

The *Pioneer* probes sent the first close-up information about Jupiter back to Earth. After completing its work, *Pioneer 10* continued to travel out into the region occupied by the outer planets, and eventually it will leave the solar system altogether. The path of *Pioneer 11* was chosen so that the probe would receive an additional impetus from its close encounter with Jupiter that would send it on in the direction of Saturn, or more accurately, the probe was put into an orbit that would take it to an encounter with Saturn five years later. From that time on *Pioneer 11* was known as *Pioneer Saturn*.

The next visits to Jupiter came from two probes in the *Voyager* series. Launched in September 1977, *Voyager 1* reached the planet on 5 March 1979. The probe penetrated as far as the region within the orbit of Io, the innermost of the moons discovered by Galileo. In July of the same year, *Voyager 2*, which had been launched somewhat earlier, arrived at Jupiter after having crossed the asteroid belt, as had its sister craft.

The *Voyager* probes provided us with the finest pictures of the cloud structure on Jupiter, close-ups of the planet, and of many of its moons. But the probes also carried counters to register energetic particles in various energy bands, instruments for measuring magnetic fields, experiments for monitoring cosmic rays, and aerials for radio-astronomy experiments.

▪ Jupiter's Own Source of Heat

From the mass of a planet and its diameter, it is possible to calculate the force of gravity at its surface. A mass of one kilogram would weigh nearly three times as much on Jupiter as it does on earth. If a man stood on the surface of Jupiter, it would feel as if two others were standing on his shoulders. However, there is no solid surface on Jupiter on which anyone could stand — but more of that later. The gravity binds nearly all the molecules of gas in the atmosphere to the planet. An upward, initial velocity of 60 kilometers per second would be required for anything to escape permanently from Jupiter. There are practically no molecules in Jupiter's atmosphere that are moving with such a high velocity. As a result, throughout its history it has lost none of its atmosphere.

The materials occurring within it are the same as those present when the planet was formed. It primarily consists of hydrogen, the same material that went to form the sun. As with the sun, other elements are present in Jupiter. Its atmosphere has been found to contain helium, water vapor, methane, and ammonia. Therefore, the planet actually reeks of ammonia. To our way of thinking the temperatures in its atmosphere are very low, about −150°C. Only lower down is it warmer. Occasionally, the opaque layer of clouds is broken, particularly in the darker cloud belts, and it is possible to see down into the lower layers. The instruments on the probes were able to measure the heat radiation coming from those parts of the atmosphere and found that the temperatures were around 0°C. The interior of Jupiter is warm; just as it becomes hotter inside the earth the deeper one goes, so Jupiter becomes warmer at greater depths. It had already been noted from earth that Jupiter emits more heat radiation than it receives from the sun. It is now thought that Jupiter radiates away about twice as much heat as it gains from the sun.

Whereas the weather on earth is governed by solar radiation, on Jupiter it is the planet's own heat that controls the processes occurring in the atmosphere. On Jupiter, the driving force for the weather comes from below. The energy that is making its way out to space is responsible for the motions of the clouds. It produces storms and hurricanes and creates and destroys individual cloud features. But the solar radiation is not completely without effects, although the size of the sun is only about one-fifth of the diameter it has here on earth. In the outermost layers of Jupiter's atmosphere, the atoms are subject to the ultraviolet radiation from the sun. Electrons are split from their parent atoms. Positively charged ions and free electrons form an outer layer, so that Jupiter's atmosphere, like the earth's, has an ionosphere, a layer that is rich in ions and electrons. Although Jupiter's ionosphere cannot be seen, it has been thoroughly investigated. As seen from Earth, both *Voyager* probes disappeared behind Jupiter for a time, during which we were unable to receive any signals from the probes. But as they vanished behind the disk, and later as they emerged on the other side, their radio waves passed through Jupiter's atmosphere on their way toward us. Thus, the probes provided "back-lighting" for Jupiter's ionosphere, since ions and electrons affect the scattering of radio waves. It was possible to investigate the ionosphere when the probe was passing onto the night side after having been subject to solar radiation during Jupiter's daytime. On the other side of the disk, the ionosphere was observed as it was again affected by solar radiation after Jupiter's night.

▪ A Bottomless Atmosphere

The materials in Jupiter's atmosphere are well known to physicists and chemists. They know the temperatures and pressures at which chemical reactions can occur within them. Knowing the force of gravity and the measured temperatures, it is possible to build a model of Jupiter's atmosphere.

Let us imagine that we are sitting in a space probe that has been launched toward Jupiter and is now falling freely down through its atmosphere. Let us assume that we begin our fall in the planet's stratosphere, where the atmospheric pressure is only one one-hundredth of the pressure that we are used to on the surface of the earth. If we were to look up toward the sky, we would see practically no clouds. Occasionally, there would be a few thin streaks of cloud in the stratosphere. The dense layers of cloud lie about 80 kilometers below us. It will take us slightly more than a minute to fall that distance. Then we find ourselves surrounded by dense clouds of ammonia crystals. The air pressure is half that at the earth's surface, and we take just a few seconds to drop through the layer of clouds. After 10 seconds, we enter a layer of snowflakes of ammonium hydrosulphide, whose molecules consist of hydrogen, nitrogen, and sulphur atoms. We drop another 10 seconds in free fall through this snow flurry, and then the clouds consist of ice crystals, like our cirrus clouds. The temperature is now close to 0°C. We might almost feel at home, but the air pressure is already five times as high as on the earth. The lower we go, the higher the temperature and pressure rise. We are now falling through a layer of fog consisting of droplets of water and ammonia.

We need have no fear of crashing into any solid planetary surface. There is none. Jupiter's atmosphere becomes denser and denser, the deeper we go, until it finally becomes a liquid. After another four-and-a-half minutes, we will splash into a sea of liquid hydrogen. Our imaginary journey in free fall through Jupiter's atmosphere would actually take much longer, because the dense atmosphere would slow down our fall very rapidly. In the forthcoming *Galileo* mission (see below), a detachable probe will be dropped into Jupiter's atmosphere with a parachute as a brake.

We know very little about the interior of Jupiter. From its mass and diameter, we can calculate its mean density, which turns out to be 1.33 grams per cubic centimeter. Thus, Jupiter is only slightly denser than liquid water. Although its dimensions and mass may appear gigantic, its density is only a quarter of the earth's mean density. It appears to consist of mainly warm, liquid hydrogen. Its central temperature is estimated to be 25,000°C. If Jupiter does consist of the same original material as went to form the sun, then it must contain the same chemical elements as are found in the sun, a few thousandths of its mass must be iron. For this reason, it is believed that there may be an iron core hidden in the center of Jupiter, with a diameter of perhaps 25,000 kilometers.

Where does Jupiter's internal energy come from? We think we know the answer now. When it formed from primordial material, the process was similar to the one that occurs when a star is formed. Masses of gas clumped together with a corresponding rise in their gravitational attraction, and as a result of their own gravity, they became denser and denser. When a body is acted upon by gravity, it gains energy. Water, which is drawn by the earth's gravity down from the mountains into rivers and then into the sea, can drive power stations. Stones that fall onto the earth from space can cause impact craters. The masses of gas that formed Jupiter, attracted by their own gravity, clumped together and gained energy. Part of this heat then went toward heating the accumulating cloud of matter.

As an incipient star becomes denser and denser under the influence of its own gravity, the temperature at its center rises above 10 million degrees Celsius. That happens to be the critical temperature for the onset of the fusion of hydrogen into helium. At that instant, the star becomes a nuclear reactor and will subsequently pass most of the remainder of its life dependent on the energy provided by hydrogen fusion. But not all masses of gas reach such a high temperature as they condense. Only masses of more than one-tenth of a solar mass reach the critical temperature of 10 million degrees Celsius. Smaller gas clouds condense without being able to draw on their potential supply of nuclear energy, because they never reach the critical temperature. Initially they heat up, but after their density has become moderately high, they begin to cool down. Jupiter is just such a failed sun. With only about one one-thousandth of a solar mass, it never had the chance of drawing on its hydrogen supply as a source of nuclear energy. The gas clouds from which it formed, however, condensed and thus became heated. Since then, Jupiter has been cooling down, and the heat is released into its atmosphere. There the flow of energy from the interior sets up rising and falling currents in the atmospheric gases. They are the type of motion that we can observe on Earth in the air above a hotplate or above an open fire.

Although with respect to mass, Jupiter may have been at a disadvantage relative to the sun when the solar system was formed, it received more than its full share of angular momentum. Even with a small telescope, it is easy to see that it is strongly flattened by its rotation. Its axis of rotation is almost perpendicular to its orbital plane. Since seasons are caused by the inclination of the rotational axis to a planet's orbital plane, as on earth (see Figure 23), there are no seasons on Jupiter. The planet's equator rotates slightly faster than its poles, where a point takes five minutes longer to complete one rotation around the axis than a point does at the equator, because there is no solid surface to force the atmosphere to rotate as a whole. The rapid rotation affects the motion of the rising and falling gas currents in the atmosphere. Nearly every current is deflected either toward the east or toward the west. This produces the banded structure, running like parallels of latitude around the planet. The rotation is not the only reason for the rapid velocities of its atmosphere; there are also individual storms. Even when moving with the speed of Jupiter's rotation, a meteorologist would measure wind speeds of up to 100 meters per second.

The Great Red Spot in the southern hemisphere is a persistent, circular storm. The *Voyager* probes determined the properties of the motions both within it and in its vicinity. When cloud centers approach it, they usually circle it several times. Other small eddies in the southern hemisphere rotate like the Great Red Spot in an counterclockwise direction; eddies in the northern hemisphere rotate clockwise. The atmosphere of Jupiter consists of nothing but systems of currents and eddies. It is an ideal exercise ground for all hydrodynamicists.

I should like to mention here a historical curiosity that may perhaps have already struck anyone reading Kepler's words quoted at the beginning of this chapter. Jupiter's spot appeared on its southern hemisphere in 1878 and since then has been called the Great Red Spot. But indications of it can be recognized on earlier drawings, perhaps even going back as far as a drawing made in 1664. That was fifty years after Kepler had written that sentence. If the spot was visible in Kepler's time, then someone must have observed Jupiter very carefully with a good telescope, which Kepler did not possess. Nevertheless, the sentence is authentic. How can this be explained?

In Chapter 5 we said that Galileo encrypted his discovery of the phases of Venus in an anagram. This cryptic text was in a letter sent to the Tuscan envoy in Prague. Kepler tried to solve the anagram by permutating the position of the letters. After a lot of tedious work, he found eight different results, which he set down in a letter to Galileo. Although these did not include the true solution, they did include the sentence quoted. It was a chance statement that was shown to be true only centuries later.

▪ The Moons of Jupiter

Galileo saw Jupiter's satellites in December 1609. Was he the first to see them? Simon Marius (1573–1624), a Bavarian astronomer, subsequently claimed that he had seen them before Galileo. Whatever the truth was, Marius was the one who proposed the names that are still used: Io, the innermost of the moons found by Galileo; then Europa, Ganymede, and Callisto. They are the largest in the list of Jupiter's family of moons, which even today is probably not complete. The diameters of the Galilean satellites are a few thousand kilometers. Their size is comparable with that of planets, and Ganymede is actually larger than Mercury. Subsequent satellites of Jupiter are much smaller. The potato-shaped moon Amalthea has already appeared in Herr Meyer's dream. Other than the Galilean satellites, it is the largest moon. Then comes Himalia with a diameter of 170 kilometers; it has been known since 1904. All the others known so far are smaller than 50 kilometers across. Metis was discovered by the *Voyager* probes and revolves around the planet inside Io's orbit in only 7 hours and 4 minutes, which is faster than Jupiter's rotation. The outermost satellites of Jupiter lie at distances of more than 20 million kilometers and their orbital periods are about two earth-years. Whereas the inner satellites revolve around the planet in the same direction as its rotation, and their orbital planes closely approximate to Jupiter's equatorial plane, the outer satellites are mere casual companions. Their orbital planes lie nowhere near the equatorial plane. The four outermost even orbit in the opposite direction to the planet's rotation. They are believed to be asteroids that at some time in the past came close to Jupiter, and which by the combined effects of the gravitational fields of Jupiter and the sun were captured in orbits around the planet.

Jupiter's ring has only been known since the *Voyager* missions. Individual lumps of rock each describe their own orbits around the planet. It is not known how much fragmented material exists in Jupiter's ring, because we do not know how large the lumps of rock are. One zone of the rings probably consists of fine dust, and another probably contains rocks ranging up to a few kilometers in size. The outer edge of the ring lies at about 55,000 kilometers above the cloud layer and is therefore in the region where the innermost moon, Metis, is orbiting. Another moon discovered on *Voyager* images, Adrastrea, also orbits within the rings, which probably reach down as far as the atmosphere of the planet, where they may mingle with the gases of the upper atmosphere.

▪ *Signals from Jupiter*

In the summer of 1954, the Department of Earth Magnetism of the renowned Carnegie Institution commissioned a new radio telescope in Florida. A total of 128 radio aerials were arranged in two sets so that they could examine the sky: 64 reflecting aerials were set in a north–south direction, and the same number in an east–west array. Seen from above, the telescope appeared like a gigantic cross. Each arm of the cross was 320 meters in length. By appropriately switching the output from the aerials, the receiver could pick up radiation from just the point that was at the zenith at any particular time. Over the course of 24 hours, a sample was taken from a narrow band around the sky, while the array of aerials rotated once because of the rotation of the earth. By changing the switching of the aerials slightly, it was possible to record radiation from strips of sky either to the south or to the north of the zenith. In the spring of 1955, two astronomers, Bernard Burke and Ken L. Franklin, were investigating a strip of sky that had been chosen to include two interesting radio sources, which were carried across the line of sight every day. One source was the Crab Nebula, the remnant of a stellar explosion that occurred in 1054, and the other was a galaxy that is a powerful radio emitter. The wavelength being used was 13.5 meters in the shortwave band, not far from the 11-meter band used by CB enthusiasts. The two radio sources could be clearly seen on the chart recordings. But there was a third source, which crossed the line of sight shortly after the Crab Nebula. Over the course of a few weeks, the two astronomers could see that this source was gradually overtaking the Crab, following it at shorter and shorter intervals.

A colleague, who had been investigating Jupiter shortly beforehand, asked if the puzzling radio source could be Jupiter. At that time, such a suggestion was thought to be absurd. People expected radiation in that shortwave region to be produced by extremely hot objects, such as flares on the sun, by hot masses of gas in space, or by mysterious galaxies far off in the depths of space. A garden-variety object like a planet could certainly not be responsible for radio waves at a length of a few meters. But the suggestion had been made, and Burke and Franklin looked up the position of Jupiter in the astronomical yearbooks. They found that the planet was indeed at the position from which the radiation was coming and that its motion corresponded with that shown by the unknown radio source. When, in the course of 1955, Jupiter reversed its direction in a loop and began to move westward again, the radio source did the same. It was perfectly clear: Jupiter was a radio source! Subsequently, it was found that the signals from Jupiter had been detected five years earlier in Australia. But no one had noticed them.

▪ Centimeter and Decimeter Wavelengths

Three years later, more was discovered about Jupiter's radio emission. Two astronomers, the American Frank D. Drake and the Swede Hein Hvatum, working at the Naval Research Laboratory of the University of Maryland, investigated Jupiter's radiation at centimeter wavelengths in order to determine the temperature of Jupiter's surface.

Every warm body emits radiation. Glowing iron emits visible light, which is radiation at wavelengths of thousandths of a millimeter. As it cools, the wavelength becomes longer. Objects at room temperature radiate at wavelengths in the region around one one-hundredth of a millimeter. Still cooler bodies mainly radiate at radio wavelengths. Radiation from warm bodies in the radio region drops sharply with decreasing temperature. So it caused considerable surprise when Drake and Hvatum found stronger radiation from Jupiter at a wavelength of 30 centimeters than at shorter wavelengths. The longer the wavelength they used, the stronger the measured radiation became. Thus, Jupiter was not only emitting radiation appropriate to its temperature, but there appeared to be another source of radiation, which became more powerful with increasing wavelength (Figure 64).

Radiation from other cosmic radio sources was known to show this sort of behavior. It originates from fast-moving electrons that are traveling through space with velocities close to that of light. If they are moving in a magnetic field, they emit radio waves whose strength increases with increasing wavelength. High-velocity electrons moving within a planetary magnetic field were already known to exist in the earth's radiation belts (see Figure 32). Thus, long before the time of the *Pioneer* probes, it was known that Jupiter had a magnetic field within which fast-moving particles were trapped.

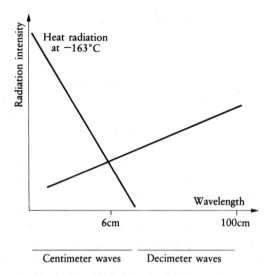

Figure 64. *A schematic diagram of the radio emission from Jupiter at wavelengths of less than one meter. The radiation consists of two components. In the centimeter region, we receive radiation emitted by a body with a temperature of 110°C above absolute zero, that is, at −163°C. The strength of this emission declines with increasing wavelength. At decimeter wavelengths, the radiation comes from electrons moving at high velocities that are trapped in Jupiter's magnetic field. The intensity of this radiation increases with increasing wavelength. The region containing decameter and longer wavelengths (see below) lies off to the right, beyond the diagram.*

When the *Pioneer* probes subsequently flew through Jupiter's radiation belts, their magnetometers measured the magnetic fields, and particle counters counted the electrons. Jupiter's magnetic field reaches far beyond the Galilean moons. It contains trapped, high-energy electrons. As *Pioneer 1* approached Jupiter, every square centimeter of the surface of the probe was hit every second by about one billion electrons, which were moving at about half the speed of light. The magnetic field is about 20,000 times stronger than that of the earth. The fast-moving electrons are responsible for the radiation at decimeter wavelengths, whereas the centimeter radiation arises because of the planet's actual temperature.

The most puzzling radiation, however, were the radio waves that were then discovered with wavelengths of more than ten meters, which are known as *decametric radiation*.

▪ *Io and the Decametric Radiation*

The type of radiation from Jupiter that was discovered by Burke and Franklin does not remain constant. It exhibits bursts in which the strength of radiation increases tenfold for a few seconds. When it was first discovered, the bursts appeared to arrive at random. Then it was noticed that they were particularly frequent when Io was either east or west of Jupiter as seen from Earth, that is, when in its circular orbit around Jupiter, it was either directly approaching or receding from the earth. What connection could there be between the direction in which Io was traveling and the radio bursts?

Being the innermost of the four major satellites of Jupiter, Io orbits where the magnetic field is particularly strong. We may expect Io's fluid interior to have high electrical conductivity. The satellite is therefore an electrical conductor that is moving in a magnetic field, like the rotor in a dynamo. Just as in a dynamo, electrical currents are set up in the satellite that produce additional electrical fields, which not only influence the electrons trapped in Jupiter's magnetic field but also produce radio waves. They mainly propagate in the direction in which the satellite is moving and in the opposite direction. When either of these directions coincides with that of the earth, we detect radio bursts from Jupiter.

In his dream, Herr Meyer saw a volcanic eruption on Io. In fact, this satellite, shown in Color Plate 5, has several volcanoes that are still active. Io is smaller than Mercury and only slightly larger than the earth's moon. Why has its interior not cooled long ago? What is the source of the energy that has kept its interior from cooling? The answer is the tidal forces acting on Io. Its orbit around Jupiter is not precisely circular, because the gravitational forces of the neighboring moons Europa and Ganymede cause pertubations in the orbit. Although tidal effects have caused its period of rotation to be generally the same as its orbital period, so that the satellite always turns the same face toward Jupiter, absolute agreement would only be achieved if the satellite's motion around Jupiter were perfectly regular. Thus, just as the moon's rotation cannot precisely match its motion along its elliptical orbit around the earth (see Chapter 4), Io only approximately approaches that condition. For an observer on Io, Jupiter would not remain motionless in the sky; its disk would move back and forth with a period of 42 hours. As a result, Io is continually deformed by tidal forces. Although it has no ocean, its interior is fluid, and its crust must respond to the tides. The two tidal bulges do not circulate around the moon as the earth's do, but instead move back and forth across the surface of Io. This continuous flexing keeps the material within Io hot and fluid. From time to time, the liquid breaks out in a volcanic eruption.

Since the gravity on Io is very low — the escape velocity is only about 2.3 kilometers per second — some of the material, which is ejected to heights of hundreds of kilometers in the form of gas and dust, escapes from the satellite and is trapped in the magnetic field, which is rotating with Jupiter. Because, like Jupiter, the field rotates once in about 10 hours while Io takes 42 hours to complete an orbit, the material flowing away from Io moves faster than the satellite. It flies away from it, and then catches up with it again. The gas that has escaped from Io collects along its orbit in the form of a tube, or torus. Both *Voyager* probes detected this tube of sulphur and other gases that have escaped from Io.

▪ *The* Galileo *Mission*

A new mission to Jupiter, its satellites, and its rings was set for May 1986. The space shuttle was due to lift a new probe into space, but the *Challenger* disaster in January 1986 set back the schedule. The launch had to be postponed until the shuttle flights were resumed. At that time, *Galileo* was brought into the waiting line again and was finally fired into orbit on 18 October 1989 by the space shuttle *Atlantis*.

A week before the launch, there were still problems. Since at Jupiter's distance from the sun the probe would not be able to collect the amount of solar energy necessary for its operation, it was equipped with two small nuclear reactors, as had already been the case with the *Voyager* probes. Each of *Galileo*'s two 280-watt batteries contains 11 kilograms of plutonium dioxide. Plutonium is produced in earthbound nuclear reactors. The nuclei of this element, which is very rare in nature — one estimate is that altogether there are only a few kilograms of natural plutonium on our planet — contain 94 protons and about 145 neutrons. The plutonium used in the reactors of the space probes is plutonium 238, the type that has only 144 neutrons in its nuclei (for the way nuclei are characterized see Note 1 to Appendix B). While it decays within its characteristic time of 88 years, the nuclei deliver energy.

Plutonium is poisonous, and consequently the announcement of *Galileo*'s launch caused many protests. A catastrophe like that of *Challenger* could easily scatter 22 kilograms of plutonium dioxide widely around the point of impact. But finally, *Galileo* was launched successfully, and three days later it left its orbit to start the long flight to Jupiter. Originally, the plan was to send the probe directly into the outer regions of the solar system. But the delay demanded drastic changes in the program. *Galileo* started its way in an inward direction. It passed Venus in the spring of 1990 to get the necessary momentum for its long journey. It will pass close to the earth in December 1990 and then again in December 1992. During both encounters with our planet, it will gain more momentum and will reach Jupiter in December 1995.

It will then fly through the region of the four Galilean satellites, passing Io at a distance of only 1000 kilometers. It will investigate Jupiter's magnetic field and finally eject a smaller probe, which, braked by a parachute, will descend into Jupiter's atmosphere and transmit data. At a depth of 100 kilometers, it will be destroyed by the high pressure. At that time, the parent ship will fire its rockets and go into an orbit around the planet.

TEN

The Lord of the Rings

For Saturn alone stands apart from the pattern of the remaining celestial bodies, and shows so many discrepant phases, that hitherto it has been doubted whether it is a globe connected to two smaller globes or whether it is a spheroid provided with two conspicuous cavities or, if you like, spots, or whether it represents a kind of vessel with handles on both sides, or finally, whether it is some other shape.

SIR CHRISTOPHER WREN (1632–1723)

■

After its encounter with Jupiter in December 1974 *Pioneer 11* was hurled in the direction of Saturn. The planet was on the opposite side of the sun. The probe was sent into an arc high above the orbital planes of the planets (Figure 65). At the beginning of 1976, it was 150 million kilometers above the plane of the earth's orbit, which corresponds to the distance between the earth and the sun. The impetus that had been given to the probe by Jupiter sent it only in the general direction of Saturn. As it approached its target, the orbit could be refined by the probe's rocket motor. But what path should be chosen? There were many ways in which it could fly past Saturn. Should the closest possible approach be chosen, somewhere between Saturn's rings and the cloud-covered atmosphere? That would not be without danger. Whatever sort of particles made up the rings, they were orbiting the planet with velocities of more than 20 kilometers per second. Any particle traveling at

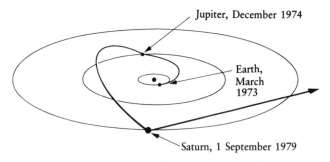

Figure 65. *The path of* Pioneer Saturn *on its way from Jupiter to Saturn rose high above the plane of the orbits of the planets. In September 1979, the probe passed only 21,000 kilometers above the outermost cloud layers of Saturn and was then flung out of the solar system.*

such a velocity would go straight through the probe. Should a flyby at a safe distance be chosen, say, far outside the outer limit of the rings? When the scientists responsible for the planned measurements took a vote, eleven were in favor of the more risky, near encounter, and only one for the safer, distant one. But science is not determined by democratic vote. In May 1978, NASA decided to alter the orbit so that *Pioneer Saturn*, as *Pioneer 11* was now called, would pass the planet at a safe distance. In July 1978, the probe's motor was started for the last time. Then the probe continued without further corrections, following the path dictated by Newtonian mechanics for an object free from any influences.

The *Voyager* probes were close to Jupiter as *Pioneer Saturn* neared its target in September 1979. Shortly after midnight on 1 September, came the first surprise. About 4000 kilometers outside the outer edge of the known rings, there proved to be yet another ring. In addition, a new, previously unknown moon was discovered.

Only a few hours later, *Pioneer Saturn* was due to fly through the ring plane, outside of the rings visible from earth it was true, but now the new ring had shown that this region of space was not necessarily empty. Might not the whole ring plane be full of lumps of rock, which only formed visible rings where they were particularly close together?

If an object in the ring plane were to destroy the probe, news would reach the earth only 86 minutes later, since light and radio waves took that long to cross the distance between the planet and the earth. Taking this delay into consideration, a signal emitted at the time of closet encounter would reach the earth at 10:28 Universal Time.

The NASA report describes the tense minutes in the control center:

As the "moment of truth" approached, a hushed silence fell over the watchers. The count-down to expected ring-plane crossing was broadcast across the nation on public radio and television, as well as throughout the sprawling Ames Research Center. At 10:28 the radio signal continued to be received. Another slow minute was ticked off in case the original estimate had been in error. Still the data flowed back from *Pioneer* and were received at the Deep Space Network station in the Mohave desert and relayed to Ames. Finally, Project Manager Charlie Hall said he thought we had made it. There were scattered cheers and many sighs of relief. A few rueful comments were also made about not having tried for the target point inside the rings. But there was little time for either celebration or regrets, since *Pioneer* was now on the most crucial leg of its journey, exploring space near Saturn that would probably not be visited by spacecraft again in this century. The closest approach to Saturn, just 21,000 kilometers above the clouds, would take place in 29 minutes. The speed of *Pioneer* at this time would reach 114,000 kilometers per hour.

▪ *The Planet with Twin Handles*

On the morning of 15 July 1610, Galileo turned his homemade telescope onto Saturn and noted that the planet's shape was completely different from that of Jupiter. On 30 July, he wrote to the Tuscan envoy to the court in Prague that Saturn was not a single star but consisted of three stars close together. However, he requested that his discovery should not be made common knowledge before he had established his priority by some form of publication. Shortly afterwards, he did just this, using an anagram. Kepler tried to find out the meaning by transposing the letters in the nonsensical

smaismrmilmepoetaleumibunenugttauiras

but in vain.

Three months later Galileo sent the solution, not to his colleague Kepler, but to the emperor: "Altissimum planetam tergeminum obseruaui," which means: "I have observed the highest planet to be triple in form." The term "highest planet" meant the outermost planet then known, in other words, Saturn. Galileo had discovered the satellites of Jupiter shortly before, which led him to believe that the object that he saw through his inadequate telescope was a body accompanied by two companions. For the next year, he kept observing Saturn. But unlike the satellites of Jupiter, which altered their positions with respect to the planet every night, the appearance of Saturn and the two secondary bodies did not seem to change. Saturn was obviously a very boring object, and Galileo did not observe it very frequently. He was in for a surprise when he again turned his telescope onto the object at the end of 1612. Saturn had suddenly become a circular object, like Jupiter, and the two companions had disappeared. Had they flown off into space, or had the planet somehow engulfed them? We have no idea how Galileo explained the change.

In subsequent years, larger telescopes were built; these were generally longer, because lenses with long focal lengths give images with smaller errors. But this exacted a price: The monster telescopes could be directed at a particular point on the sky only with great difficulty, and they waved about in the slightest breeze. Hevelius (see p. 87) in Danzig worked with a giant telescope 45 meters long. In Holland, Christiaan Huygens (see p. 122) had, among his collection of telescopes, one 64 meters long. Later, when it was learnt how to manufacture better lenses and combinations of lenses, such unwieldy, monster instruments became unnecessary.

The uncertainty that was then felt about Saturn's true shape is shown by the quotation at the beginning of this chapter, written by Newton's friend Sir Christopher Wren, who is primarily known as the architect of St. Paul's Cathedral in London. The next decisive step was made by the Dutchman Christiaan Huygens. On 25 March 1655, he saw a small star near Saturn. Observations over following nights showed that Saturn had a satellite. Huygens had discovered Titan, the largest and brightest of Saturn's moons. Both Wren and Hevelius had previously seen the point of light, but both had taken it for a fixed star. Then Huygens saw the solution to the riddle of Saturn's strange shape, but initially he too resorted to using a cryptic anagram to establish his scientific priority. He published the solution only three years later. "It is surrounded by a thin, flat ring, nowhere touching, and inclined to the ecliptic." (The ecliptic is the earth's orbital plane.) The inclination to the ecliptic allowed Huygens to account not only for the planet's appearance through a telescope, but also for the disappearance of Galileo's "secondary spheres." Figure 66 shows Huygen's explanation of the variation in the appearance of Saturn over the course of a year. His theory about Saturn's ring was not generally accepted by his colleagues as the long-awaited solution to the problem. Hevelius, for example, took umbrage over the fact that Huygens considered his own telescopes to be superior.

Around 1660, Huygens was one of the best astronomical observers of his day. Others then came with better telescopes, but they only confirmed Huygens's Saturn-ring theory. Huygens predicted that in the summer of 1671, Saturn's ring would no longer be visible from Earth, but should reappear the following summer. His predictions were fulfilled, confirming his model.

In 1675, Cassini, whom we already mentioned, saw a dark circular line in Saturn's ring, which was a division between two zones of the rings. The outer ring, which is now known as Ring A, surrounds the inner rings and the planet. Separated from it by the Cassini Division is the inner ring (Ring B).

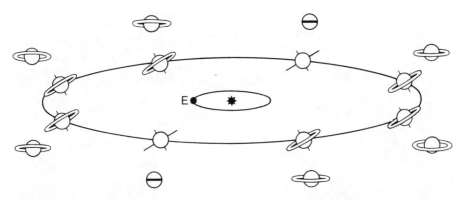

Figure 66. *The explanation of the changing appearance of Saturn's ring, according to Christiaan Huygens. The orbits of the earth and Saturn around the sun are shown. Outside the orbit of Saturn, the appearance of Saturn's ring, as seen by observers on the earth (E) is indicated. Over the course of one of Saturn's years, on two occasions they will see the ring, which does not change its orientation, from the side, when it will be invisible. During the intervening periods, it will either be seen at an angle "from above" or "from below."*

Farther inside, another, transparent ring was subsequently recognized, indicated in Figure 67 as Ring C (see also Color Plate 6).

Rings around a planet must rotate, because only then can their centrifugal force counteract their tendency to collapse down onto the planet. The rings consist of individual lumps of rock, which move around Saturn in a way similar to that in which the planets orbit the sun. They also obey Kepler's third law (see Chapter 1), and therefore, the lumps of rock in the outermost orbits require a longer time to orbit the planet than do those closer in.

With the improvements in telescopes, additional satellites of Saturn were discovered. Cassini found Iapetus, Rhea, Tethys, and Dione. William Herschel (see Chapter 4) found Mimas and Enceladus. He also determined the planet's rotational period as 10 hours and 16 minutes—only two minutes longer than the value accepted nowadays. In the middle of the last century, a watchmaker from Boston, William Cranch Bond (1789–1859), discovered the eighth satellite of Saturn, Hyperion; and William Pickering (1858–1938) of the Harvard College Observatory photographed the ninth satellite, Phoebe, in 1898. We now know of more than twenty satellites of Saturn.

Saturn's orbit is nearly twice the size of Jupiter's. As a result, Saturn requires 10,759.20 earth-days—nearly 29.5 years—to complete its orbit. Again this is in accordance with Kepler's third law (see Chapter 1). During the course of an earth-year, it only moves a short distance along its orbit; thus, oppositions of Saturn occur about 13 days later each year. The phases of the ring system recur every 29.5 years. For roughly half of that period we see the

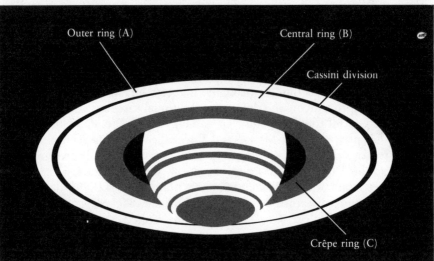

Figure 67. *Three of Saturn's rings can be seen from Earth. The photograph (top) and the schematic drawing (bottom) show the outer A Ring and, inside it, the B Ring. Between the two lies the dark strip of the Cassini Division. Even farther in, but not visible on the photograph, is the faint, inner Crêpe ring (Photo: Palomar Observatory).*

rings only from the north, and for the remainder only from the south. In between times, we do not see them at all, because they cannot be detected when viewed from the side.

Saturn rotates somewhat more slowly than Jupiter, one day at Saturn's equator being equivalent to 10 hours and 14 minutes. Rotation flattens a planet, working against gravity, which tries to force it to become a sphere. On Saturn, the gravity is less than half that on Jupiter. Thus, although centrifugal force is weaker on Saturn, it causes a greater flattening of the planet than occurs on Jupiter. Saturn's rotational axis is inclined to its orbital plane like that of the earth, and so during the (30-earth-year) Saturnian year, both the northern and southern hemispheres experience summer and winter. That does not make much of a difference, however, because the apparent diameter of the sun is only about one-tenth of the size as seen from earth. Like Jupiter, Saturn's atmosphere receives more heat from its interior than it does from the sun.

Gallia Passes Through Saturn's Ring

When Herr Meyer, in another dream, found himself on Gallia again, the sky was pitch-black. The ice in the cometary nucleus had ceased to evaporate. Dust was no longer able to escape into space, and the jet on the horizon had disappeared. The light from a bright star that had just risen cast shadows of the measuring instruments at the window across onto the opposite wall.

"Seen from Saturn's distance, the sun looks almost like a point of light, doesn't it?" said Professor Rosette, who had come up behind Herr Meyer. "But never mind the sun," he continued, "in an hour, Gallia will pass through Saturn's rings. We shall go straight through its Cassini Division. Luckily, we can expect fewer lumps of rock there than in either Ring A or Ring B. Nevertheless, the most dangerous stage of our journey is probably approaching."

Then the ring rose above the horizon. It was a giant disk, consisting of innumerable individual circles, looking like the grooves in a record (Figure 68). When Herr Meyer looked through his binoculars, even the narrowest lines visible to the naked eye proved to consist of yet finer lines.

"Can you see the outermost ring, the F Ring? It was discovered by *Pioneer Saturn*." With some difficulty Herr Meyer made out an extremely faint ring right outside the others; it consisted of numerous, even finer, concentric rings.

"Look at the two satellites of Saturn that hold the F Ring in place with their gravitational effects. One of them orbits just inside

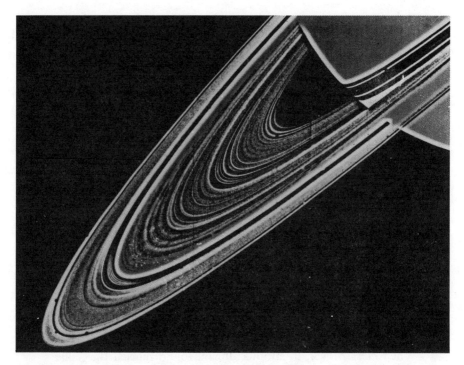

Figure 68. *In close-up images, Saturn's rings resolve into countless individual rings (Photo: NASA/JPL).*

the tiny ring, and the other just outside." Herr Meyer looked in both of the places that Rosette had pointed out. Through his binoculars, he could see an irregularly shaped object at each of the positions.

"If one of the ring particles attempts to move outward, it is slowed down by the gravitational force of the outer, slower-moving moon. Its centrifugal force becomes less, and Saturn's gravity pulls it back into the ring. If one of the particles attempts to move inward, it is affected by the inner, more rapidly moving moon. Its orbital motion is speeded up, and centrifugal force pushes it back out into position. The two satellites act like two sheepdogs, herding the rocks in the F Ring, and are therefore also known as shepherd satellites."

But Herr Meyer was not really listening. He was staring at the broad ring that had now risen well above the horizon. Its surface was very bright, and it consisted of innumerable narrow ringlets.

Then the dark band of the Cassini Division appeared. It separated the A Ring, which had risen earlier, from the B Ring, which was just beginning to rise above the horizon. Again it consisted of thousands of parallel lines of different brightness. In places there were zones where ringlets were darker than elsewhere (Figure 69). Then the disk of the planet itself rose. Herr Meyer could see very

Figure 69. *Saturn's B Ring (cf. Figure 68) again resolves into individual ringlets in this* Voyager 2 *picture. In this part of the ring system, sporadic features, which can be seen here as dark radial bands, appear and disappear. This phenomenon is surprising, because we have to assume that we are looking at individual rings that have differing orbital velocities. According to Newtonian mechanics, the pieces of ice and rock that are orbiting in each ringlet should not influence the lumps in the neighboring rings. But these features represent a phenomenon where, over a wide area of the ring system, the individual rings have some collective properties. It also seems that the particles in the dark areas are only loosely distributed (Photo: NASA/JPL).*

few features in the layer of clouds, even through the binoculars. A few spots reminded him of the cyclonic storms in Jupiter's atmosphere. But here, on Saturn, everything was much more subdued.

"Saturn is far quieter," thought Herr Meyer. When he looked at the Cassini Division again, he could see the disk of the planet through it. The apparently dark band was transparent. Even the inner portion of the B Ring seemed to be almost transparent, and through it, one could see the dark night sky and the surface of the planet.

Herr Meyer concentrated on the planet's disk. It was so large in the sky that he could just cover it with his outstretched hand. Saturn was distinctly flattened. Cloud belts ran parallel to the equator. But they showed far less contrast than those on Jupiter.

"Have you seen Titan yet?" asked the professor. Herr Meyer tore his eyes away from the planet and scanned the dark sky. Then he saw a tiny circular disk, far outside the system of rings.

"Its diameter of 5000 kilometers means that it is rather larger than Mercury," said Rosette. "Its gravity is considerably less, but it still has a dense atmosphere that contains about ten times the mass of the earth's atmosphere. It would seem that at the low temperatures that prevail on Titan, the gases cannot escape. Where daytime temperatures are of the order of $-180°C$, atoms in the air have the greatest difficulty in reaching escape velocity. Titan is the only satellite yet known to have a dense atmosphere. It is impossible to see down to the surface. Some time a probe equipped with radar will have to be put in orbit around it."

"Or else, there will have to be a manned landing," said Herr Meyer.

"I'm not sure that I would want to be there," replied Rosette. "Titan's atmosphere consists mainly of nitrogen, and oxygen is more or less completely absent. We suspect that Titan's interior consists of frozen water. All the oxygen is probably bound up in water molecules. The atmosphere therefore contains methane and other hydrocarbons. When it rains, it probably rains gasoline. In addition, prussic acid has been detected in the atmosphere."

In the meantime, the rings had come nearer, and Saturn's disk was larger. It now took two outstretched hands to cover it. Gallia was approaching at an angle to the ring, which now covered the whole of the sky visible through the window. The Cassini Division lay like a dark gulf in front of them. Herr Meyer could see stars in the dark sky beyond it. The nearer they came, the more the main rings became divided into finer and finer ringlets.

"I have compared the orbits of all known satellites of Saturn with Gallia's," said Professor Rosette. "We shall always be at a safe distance from any of them. But one can never be quite sure. When the *Pioneer Saturn* probe passed close to the plane of the rings, the counters on board measured the number of charged particles in that region of space. Suddenly, for 8 seconds, the rates recorded by several instruments dropped to zero. At the same time, the instruments recorded a magnetic disturbance. The probe had apparently passed extremely close to some object that had intercepted the flux of charged particles and had affected the magnetometers. From the duration of the disturbance, it was deduced that the diameter of the body must be about 200 kilometers. It had been realized the day before that there were previously unknown bodies near Saturn, when the probe discovered an unknown satellite that did, in fact, have a diameter of 200 kilometers. A quick calculation showed that in the intervening time, the newly found satellite had completed one orbit and was in the immediate vicinity of the probe. Apparently, *Pioneer Saturn* had just missed being hit by the satellite it had discovered the day before! I hope we are lucky!"

Through his binoculars Herr Meyer could now see that the rings consisted of individual fragments, which were all following the same orbit around the planet. The Cassini Division had now become very broad. But to his horror, Herr Meyer saw that within the division, there were ringlets where individual lumps of material followed one another along several distinct orbits.

"This division is not completely empty," said Professor Rosette. "The *Voyager* probes discovered several ringlets within it. I should not be at all surprised if we collide with an iceberg. We know that the individual fragments in the rings, which appear as lumps of rock in your binoculars, actually seem to consist of a mixture of rock and ordinary ice."

The closer they came to the ring system, the easier it was to see the individual blocks of ice, even with the naked eye. The dark Cassini Division was now even wider.

Suddenly they were in the midst of a storm of ice. The pieces all appeared to come from a single point in Gallia's sky, just like a shower of meteors on Earth. Each one of them appeared as a tiny, slowly moving speck, but as they grew, they moved too fast for the eye to follow. A few of the blocks fell on Gallia, vanishing without a trace against the ground. The impacts often sent up dark clumps of material, which disappeared into space. Then the bombardment stopped: Gallia was out of the danger zone. Herr Meyer sighed with relief.

▪ *The* Voyager *Probes*

When the *Voyager* probes reached Saturn in 1980 and 1981 after their Jupiter mission, they passed much farther from the planet than *Pioneer Saturn* the year before (Figure 70). Despite this, they radioed far more information back to Earth.

Voyager 1 arrived in November 1980. Before passing near the planet, the probe encountered Titan. It passed only 4000 kilometers above the top of the cloud layer that surrounds Saturn's largest satellite. As seen from Earth, the probe was hidden by Titan. As it passed behind the satellite, its radio waves "backlit" Titan's atmosphere. Observations from Earth showed how the signals weakened and finally disappeared behind Titan, giving a lot of information about the satellite's atmosphere.

The probe continued on toward the planet, but stayed at a safe distance from the ice in the rings. The satellites Tethys, Mimas, Enceladus, and Rhea were photographed. Tethys has some craters, but most noticeable is a rift valley about 100 kilometers wide. Rhea (Figure 71) is covered with craters and resembles the earth's moon and Mercury. Dione is also like the moon.

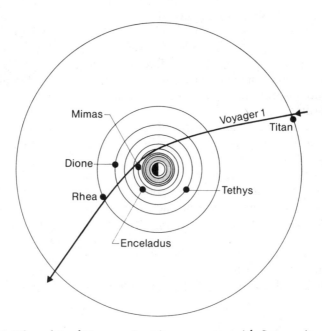

Figure 70. *The orbit of* Voyager 1 *at its encounter with Saturn, its ring system, and its satellites.*

Figure 71. *Saturn's moon Rhea has a diameter of 1530 kilometers. Its surface is saturated with impact craters. On this image, obtained by* Voyager 1, *individual features down to 1 kilometer in size can be recognized.*

In August 1981, *Voyager 2* reached Saturn and again photographed the rings and the satellites, including the second-farthest satellite, Iapetus. One image of Tethys shows a gigantic crater, 400 kilometers in diameter, nearly half the size of the satellite itself. At its closest approach to Saturn, *Voyager 2* was only 120,000 kilometers from the moon Enceladus, which proved to have both impact craters and broad crater-free areas (Figure 72). We must assume that beneath a solid crust of ice, there is liquid water that has occasionally escaped onto the surface, flooding whole areas of impact craters and subsequently freezing. It is not known why the satellite is not frozen solid. It seems to have some source of energy that prevents the water in its interior from freezing.

After the images of Enceladus had been taken. *Voyager 2* prepared to fly behind Saturn. Between 22:26 and 0:01 the probe would be behind the planet, crossing the plane of the rings. As it disappeared behind the planet and subsequently reappeared, the radio signal continuously transmitted by the probe would pass through Saturn's atmosphere. As anticipated, the signal disappeared at about 22:26. About 18 minutes after disappearing behind the

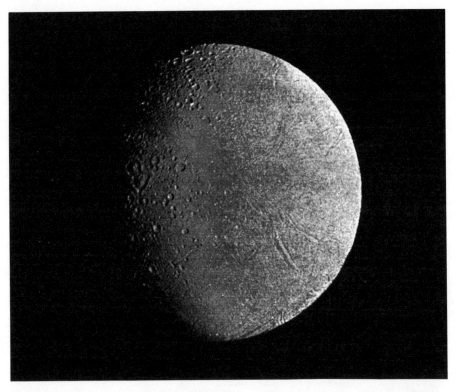

Figure 72. *An image of Saturn's satellite Enceladus, taken by* Voyager 2 *in August 1981. Its diameter is 500 kilometers, so it is even smaller than Rhea (Figure 71). In addition to impact craters, long furrows and ridges are noticeable that stretch across large areas of the surface. In contrast to the craters, these features indicate that the satellite was formed over a long enough period for the folds to be created (Photo: NASA/JPL).*

disk, the probe was due to cross the ring plane. News of anything that might happen as it did so would only be known when the probe reappeared from behind Saturn. Two minutes before midnight, the first faint signals were detected. The probe had survived! But it was soon discovered that something was wrong. The instruments were not pointing in the correct direction. While mission control on earth worked feverishly to correct the problem by suitable maneuvers, the probe flew, blind, past Enceladus and Tethys. It took several days before everything was again under control. By then, the probe was far away from Saturn, its rings, and its satellites.

▪ *The New Saturn*

The *Voyager* probes brought many surprises. The fine structure in the rings was probably the greatest, but individual features were seen that remain unexplained to this day. There are dark areas that stretch across the rings like spokes of a wheel (Figure 69). Since every body in the rings follows its own orbit around the planet, those closer to the planet orbit faster than those lying father out. Any radial structure will soon be deformed into a spiral, which becomes tighter and tighter. How then, do the "spokes" arise? They occur where the orbital period of the ring particles is the same as the planet's period of rotation. Seen from the surface of Saturn, the particles would always stay at the same point in the sky, just as geostationary satellites do when seen from Earth (see Chapter 3). Saturn has a magnetic field and as the planet rotates the field rotates in space. The same field lines always pass through the particles found in the zones where the spokes occur. Any fine dust particles in the ring that became electrically charged would not only be subject to gravitational and centrifugal forces, but to magnetic forces as well. It is suspected that Saturn's magnetism is responsible for the spokelike features.

The probes investigated the atmospheric chemistry. Saturn is a smaller version of Jupiter. Hydrogen and helium are its principal components, but ammonia, methane, and other hydrocarbons are also found, as well as a compound of hydrogen and phosphorus. The probes also detected radio bursts coming from the atmosphere, which appear to have been caused by lightning.

Saturn is similar to Jupiter in other respects as well. Beneath its atmosphere lies an ocean consisting of liquid hydrogen. Farther in, the hydrogen becomes so dense that its properties resemble those of a metal; its electrical conductivity is very high, for example. At the center, there is probably a core of heavier chemical elements; this core probably consists of rock and iron, and its mass is perhaps three earth-masses.

Saturn appears to have preserved its own interior heat, like Jupiter. It, too, is a failed sun. Its heat is a result of its original contraction. Whereas the temperature at the center of Jupiter is estimated to be 25,000°C, that at the center of Saturn is probably only 15,000°C. Because of the lesser amount of solar radiation, the amount of energy provided by the heat from the planet's interior is relatively more significant than it is on Jupiter.

Saturn's magnetic field is not as strong as Jupiter's, but it is about one thousand times as strong as the earth's. This is another sign that most of the planet's interior is fluid.

As with the earth and Jupiter, electrically charged particles are trapped in its magnetic field. Some of these originate in the sun; others come from Titan and the other satellites, because many of the moons orbit inside the radiation belts, which reach out as far as Titan. In the inner regions of the belts, many of the particles are swept up by the moons, particularly by the rings. As a probe approaches the ring system, its counters fail to detect any particles as soon as the outer rings are reached.

Like Jupiter, Saturn also appears to emit radio waves at decameter wavelengths from its magnetosphere. Although they are too weak to be detected from Earth, they have been detected by the space probes. Just as Io is responsible for the bursts from Jupiter, Saturn's satellite Mimas appears to play a part.

Within Saturn's radiation belts, temperatures of hundreds of millions of degrees were measured. That is higher than the temperature at the center of the sun. But quoting these high temperatures may be misleading. Probes have passed through these hot regions without suffering any damage, although at temperatures of millions of degrees, all known substances are vaporized. The probes survived these high temperatures because the density of the hot gases was extremely low. In fact, the probes were repeatedly hit by atoms and electrons moving at exceptionally high velocities. But when particles arrive singly, a probe has time to reradiate any heat deposited by the particle back into space. The probe itself does not acquire a temperature of millions of degrees, and its increase in temperature is probably hardly measurable. It would be quite different if the gases had a high density. Then so many particles would deliver up their heat to the probe every second that it would acquire more thermal energy than it could radiate away to space, and it would be bound to heat up. Even in the former case, however, a person would be harmed by the intense particle radiation. It would not be the high temperature of the surrounding gas that would do the damage; a person would not be heated any more than a space probe if the particles were to arrive singly. But, they would travel straight through the person's body, causing changes in individual atoms and molecules similar to the effects of nuclear radiation.

▪ Thousands of Rings

The *Voyager* images resolved Saturn's system of rings into ringlets. We can detect them down to widths of only 100 kilometers. If we were able to examine one of these ringlets even closer, we would probably find that it also consists of thousands of finer ringlets, down to perhaps one kilometer across. We do not really know why the individual blocks of ice have this tendency to occur in these narrow ringlets, although in many cases, satellites are responsible for creating and maintaining these features. We saw in Herr

Meyer's dream that the F Ring is controlled by two shepherd satellites. Another "shepherd," first discovered in 1980, orbits a few hundred kilometers outside the outer edge of the A Ring, thus preventing the lumps of ice within the ring from migrating outward.

The Cassini Division appears to be linked somehow to Saturn's moon Mimas. Bodies in the Division have an orbital period that is half that of Mimas. At some time in the past did Mimas gravitational effects cause the gap, the Cassini Division, to be cleared of particles? The region is not completely empty however; as we have seen, it does contain some ringlets. Nevertheless, there are far fewer fragments in the Division than there are in other regions of the rings.

In addition, the satellites of Saturn illustrate, on a small scale, the mathematical solution to the three-body problem. To recap, the Trojan asteroids form an equilateral triangle with the Sun and Jupiter and orbit the sun with the same period as Jupiter (see Chapter 2). Saturn's satellite Tethys similarly governs the orbits of two smaller, Trojan satellites. In 1980, two satellites were discovered that have the same orbital period as Tethys and that form equilateral triangles with it and the center of Saturn. Dione, too, has an associated Trojan satellite, also known since 1980.

The *Voyager* probes left Saturn, the planet with the thousands of rings, with a magnificent scientific booty. Their orbits now took them out to the farthest reaches of the solar system. In particular, *Voyager 2* was to encounter Uranus in January 1986 and Neptune in August 1989.

ELEVEN

The Outermost Planets

Pluto: Extreme, doubled polarization, destruction and change, purification, regeneration, psychic, magic, ambition, strong determination, inclination to artistic activity, talent for research, progressive attitude, sense of community (generation planet).

THE ASTROLOGER ELIZABETH TEISSIER

■

One outer planet was first seen by a musician; another by a Berlin astronomer; the third was found by the son of a farmer from Kansas and was given its name by an 11-year-old schoolgirl.

■ *The Bath Organist's Planet*

The 800-kilogram object had been traveling for nearly five years. Its initial impetus and the sun's gravity would determine where it ended up. Its movement had been determined by Saturn's gravitational field in August 1981.

Since then, it had been flying undisturbed through empty space, only occasionally being roused from its slumber by radio messages from Earth.

Even before *Voyager 2* encountered Jupiter in the summer of 1979, there had been problems with the radio link. After the Saturn encounter, the mechanism of the camera platform had gone on strike (see Chapter 10). It is not easy to repair equipment when it is so far away that a radio signal takes more than an hour to reach it. But the scientists and technicians at the Jet Propulsion Laboratory in Pasadena, California, had by now learned to cope with the sometimes less-than-perfect communications over extreme distances, loading new programs into the on-board computer by radio, and persuading immobile bearings to function again.

In January 1986, *Voyager 2* had reached Uranus, the first of the planets that were completely unknown to the ancients and to both Kepler and Galileo. Yet the closest of them is occasionally bright enough to be just visible to the naked eye.

Five planets had been known from antiquity. They could be seen by the unaided eye. The fact that the earth should also be counted among them had been known since the time of Copernicus. Not until the second half of the eighteenth century did people learn that there was at least one other major body orbiting the sun. The great moment came on 13 March 1781 between 10 and 11 o'clock in the evening, English time, when William Herschel (see Chapter 4), an organist at Bath, was using his homemade reflecting telescope to observe the sky near the constellation of Taurus. Close to the star Zeta Tauri, he saw a faint point of light, which he took to be a comet. On 17 March, he looked for the object again and found that in the meantime it had moved relative to the fixed stars. He followed the object over the next two months and saw it move into the constellation of Gemini.

Herschel thought that the object must be a comet, reported his observations to the astronomers at Greenwich and Oxford, and did not bother much more about it. He was not particularly interested in comets. In the next few months, many astronomers followed the object. Although most of them felt, like Herschel, that it was a comet, the idea began to be mooted that it might be a planet. News of the celestial body discovered by the Bath organist soon reached the rest of Europe.

The object was moving too slowly across the sky to be a comet. Since the small portion of its orbit that had been observed closely approximated a circle, an ellipse, or a parabola, it was impossible to determine whether this new celestial body was moving on an elongated ellipse, like a comet, or in an orbit similar to a circle, like a planet. But one thing quickly became clear: If it was a comet, then the closest point of its orbit to the sun was very distant, beyond even the orbit of Saturn. That might be why the comet did not show any sign of a tail.

Johann Elert Bode (see Chapter 4) was convinced from the start that Herschel's object was a planet. He even chose its name *Uranus*.

The farther Herschel's star moved across the sky, the better its orbit in space could be determined. It soon became clear that it was a planet, orbiting the sun outside the orbit of Saturn. In November of that year, the Royal Society awarded Herschel the Copley Medal, their highest honor, for his discovery. In the exchange of letters that accompanied the award, Herschel wrote:

> "This new star could not have been found out even with the best telescopes had I not undertaken to examine every star in the heav'ns including such as are telescopic, to the amount of at least 8 or 10 thousand. I found it at the end of my second review after a number of observations not less than 15 thousand; so that the discovery cannot be said to be owing to chance only it being almost impossible that such a star could escape my notice. . . .

At opposition, the planet can even be seen with the naked eye, but no one had ever noticed it. Because of his discovery, Herschel was elected a Fellow of the Royal Society, and on 20 May 1782, he was received by the king. Shortly afterwards, Herschel showed George III and the rest of the royal family Jupiter, Saturn, and other celestial objects through his telescope. The king begged Herschel to give up his occupation as a musician, move to Windsor, and devote his time thenceforth to astronomy. Herschel accepted the king's offer and was granted a yearly pension of 200 pounds. Herschel, the amateur, had turned professional.

It was some time before astronomers agreed on a name for the new planet. Herschel wanted to call it "Georgium Sidus" ("Georgian Star") after the king. "Herschel," "Hypercronius," "Cybele," "Astraea," and "Minerva" were other suggestions. But finally Bode's "Uranus" won the day. By the end of his life, William Herschel, the Hannoverian regimental musician who emigrated to England, had made many other exceptionally important contributions to astronomy. The telescopes that he built were of the finest of their day. In 1787, he discovered two satellites of Uranus, which were later named Titania and Oberon.

Saturn is about 10 times as far from the sun as the earth is, but Uranus orbits the sun at about 19 times the earth's distance. As a result, every square centimeter of its surface receives only about $1/400$ of the light and heat falling on a similar surface on the earth. The world out there is cold and dark. The sun does look like a dazzlingly bright star in the sky of Uranus, but it would never appear other than a brilliant point of light to our eyes, about the size of a dime viewed at a distance of 33 meters. Uranus requires 84 earth-years for one orbit, so it has only completed about two-and-a-half orbits since its discovery. In the year 2033, it will again be in the area where it was when Herschel discovered it. He noted immediately that it appeared as a small disk when seen through his telescope, about $1/450$ of the size of the full moon. No features could be seen on its disk, not even in 1972, when it was studied from a balloon.

This was the second phase of a balloon program known as Stratoscope that had been started by the Princeton astronomer Martin Schwarzschild. The program involved lifting a telescope in the gondola of a balloon into the upper atmosphere in order to photograph celestial objects from a height at which clouds, haze, and air turbulence would not affect the seeing. It had already photographed features on the surface of the sun that could only be seen very indistinctly from the ground. In 1972, Uranus was on the program of observations. Photographs taken with a reflecting telescope flying at a height of 21 kilometers enabled the Stratoscope team to determine the diameter of Uranus. Its size of 52,000 kilometers is about four times that of the earth.

No features could be detected on the disk of Uranus. That was not just a result of the great distance, since we should still be able to make out the cloud belts on Jupiter if it were at a distance where it appeared just as small. The atmosphere of Uranus is indeed featureless. Even when *Voyager 2* flew past the planet, very few features could be detected. The Stratoscope investigation of Uranus was directed by Schwarzschild's student, Robert Danielson (1931–1976). Only after his death was a sensational feature discovered on his photographs, which had been completely overlooked until then. We shall return to this later, on p. 219. The lack of detail made it very difficult to determine this planet's remarkable rotation. A first indication of it was provided by the planet's satellites.

▪ *The Satellites of Uranus*

As we have already seen, Herschel discovered two satellites in 1787. He noted that their orbits were inclined at a great angle to the plane of Uranus's orbit. William Lassell (1799–1880), an English amateur astronomer from Liverpool, found two more moons in 1851, Ariel and Umbriel. It was 1948 before the Dutch-born American astronomer Gerard P. Kuiper (1905–1973) discovered the satellite Miranda. Herschel had announced four other satellites of Uranus in 1797, but since neither he nor anyone else saw them again, he was probably mistaken.

The motions of the satellites of Uranus allowed the planet's gravity and mass to be determined. Its mass of 14.5 earth-masses is only one twenty-second of the mass of Jupiter, but it is still the fourth planet in order of mass; Jupiter, Saturn, and Neptune are all more massive. The motion of the moons, however, soon called attention to another remarkable property of Uranus.

Satellites generally reflect the rotation of their parent planet. They revolve around it in the same direction as the parent's rotation, and their orbital planes do not normally differ markedly from the plane of the equator. Only outer satellites tend to be exceptions, like the outermost moons of Jupiter. But they were most probably captured at a comparatively late date. The earth–moon system is actually exceptional. Although the moon does orbit in the same direction as the earth's rotation, its orbital plane is inclined to the plane of the earth's equator by an angle that varies between 18 and 28 degrees, depending on the rotation of the moon's line of nodes (see Figure 33). We have already discussed the anomalous properties of the moon. "Proper" planetary companions, like the Galilean satellites of Jupiter or the seven inner satellites of Saturn, lie exactly in the equatorial plane of their planet. If that were also true for the satellites of Uranus, then Uranus would have to rotate with its axis lying more or less exactly in its orbital plane.

This can actually be detected by accurate investigation of the light received from Uranus. Because of the remarkable inclination of its rotational axis, during the course of one Uranus-year, which lasts 84 earth-years, the north and south poles are at times turned toward the sun (Figure 73). Uranus pointed one of its poles directly toward the earth in 1902, 1944, and 1985.

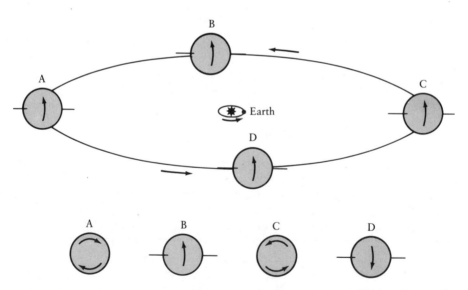

Figure 73. *The rotation of Uranus and its orbit around the sun shown schematically and not to scale. The rotation axis lies very nearly in the orbital plane. The lower portion of the diagram shows how the planet appears from Earth. At positions A and C, one hemisphere is constantly in sunlight throughout the daily rotation; the other remains in darkness. At positions B and D, the sun rises and sets each day for every point on the planet's surface.*

At such times, only one hemisphere can be seen from earth. In the interim, in 1923 and 1965 for example, we see the planet from the side. Its equator crosses the visible half of the planet, and during the course of one Uranus-day, its entire surface rotates in front of our eyes.

▪ *Unknown Uranus*

The sunlight reflected by Uranus is greenish. Closer investigation shows that the atmosphere primarily consists of hydrogen and helium, as well as small quantities of methane. The temperature is around −170°C. From its mass and diameter, the mean density of the material within Uranus could be determined. With 1.23 grams per cubic centimeter, it is comparable with that of Jupiter, denser than Saturn, but not as dense as the four inner planets. It is thought to have a solid core, with a liquid mantle of water, methane, and ammonia. Heat radiation from Uranus can be detected on Earth in the radio region. This was how variations in Uranus's atmosphere with time were detected in the 1970s; they probably originate in seasonal changes on the planet.

More was learned about the atmosphere of Uranus when the planet passed in front of a background star and eclipsed it for a short time. Before the star was hidden by the disk of the planet, its light passed successively through different layers of the atmosphere. The same happened on the other side. While it was planned to send space probes behind the planet when their radio signals would pass through the atmosphere, nature itself set up a similar, completely free experiment.

There seemed to be a possibility of this happening on 10 March 1977. It was hoped not only that it would be possible to discover something about the atmosphere of Uranus on that date, but also that the times of the beginning and end of the stellar occultation would provide information about the still uncertain diameter of the planet, and about its degree of flattening. The observation of this event brought one of the major surprises in recent planetary astronomy. It is quite a long story.

▪ *The Discovery of the Rings*

Herschel, the man who discovered Uranus, claimed in 1787 and 1789 that he had seen two rings. By 1798, however, he was convinced that Uranus did not posses rings resembling those of Saturn. Since that time, this was the general view among astronomers. But the stellar occultation by Uranus on 10 March 1977 taught us better.

The star known as SAO 158689 in the constellation of Libra can be seen with small telescopes. In 1973, an astronomer at the Royal Greenwich Observatory, Gordon Taylor, discovered that Uranus would pass in front of this faint star. This would offer the opportunity of finding out something about Uranus.

Before we come to the remarkable event of 10 March 1977, we need to consider what actually happens when a planet comes between a star and the earth, hiding the star. It casts its "shadow" on the earth, rather as the moon casts its shadow on the earth when it hides the sun's disk during a solar eclipse. Such events are difficult to predict in detail, and when it became apparent that an event might occur in 1977, it was uncertain how much of the earth would fall within the shadow. As seen from the northern hemisphere, Uranus would pass very close to the star, but a full occultation might occur somewhere over the southern hemisphere. In order to obtain better predictions, the positions of both the planet and the star had to be redetermined.

Unfortunately, these improved positions made the occultation seem even less likely to happen than had first been assumed. It looked as though the shadow might miss the earth entirely. One month before the event, it was still uncertain whether the eclipse would actually occur. But in the meantime, preparations for monitoring the event had gone ahead. On 10 March, many groups of observers had made their preparations. Workers at the observatory at Perth in Australia intended to measure the combined light of Uranus and the star. Since it was uncertain when the occultation would actually begin, observations measuring the combined light were started an hour before the predicted time. At the moment of occultation, the measured intensity of the light would drop, because the light from the star would be blocked, and only that from Uranus would contribute to the combined light.

At the same time, a Lockheed C-141 aircraft that had taken off from Perth some hours before was flying over the Pacific. This flying observatory is named the Kuiper Airborne Observatory (KAO) after the planetary scientist Gerard Kuiper, the discoverer of the fifth satellite of Uranus. It was scheduled to observe the event from a height of more than 12 kilometers, and it, too, would measure the combined light of Uranus and the star. Since Uranus is normally much brighter than the star, a wavelength was used in which most of the sunlight falling on Uranus is absorbed by methane in its atmosphere. As a result, at that wavelength, the star and Uranus appeared approximately equally as bright. When the star vanished behind Uranus, the intensity of the combined light would fall by about half. About 35 minutes before the expected time of occultation, the star became distinctly fainter for a few seconds, and four minutes later the same thing happened twice more. Then the star again vanished twice for about one second each time. The observers in the aircraft followed the combined light from the star and the planet until the eclipse began. The star was behind the disk of the planet for 25 minutes, then it reappeared, but again the light from the star dipped five times. Since the KAO can only stay in the air for 12½ hours, it was time to return.

In the plane they tried to warn observers in South Africa to follow the star and planet for as long as possible after the occultation. The leader of the group, James Elliot, sent a telegram to the Central Bureau of the International Astronomical Union, announcing that there must be many tiny bodies close to Uranus, several of which had eclipsed the background star for a few seconds. The news soon went round the world.

If the rules of the game had been the same in the aircraft as they had been in the days of Galileo and Huygens, the discovery would have been described in a Latin sentence and then encrypted in an anagram. But nowadays things are different. Anyone who wants to establish priority sends a telegram, not an anagram.

The group in Perth had observed the same events. The astronomers in Cape Town confirmed the observation. It was soon found that the dimmings before and after the main occultation did not fit the picture of many individual satellites, but indicated the presence of several rings. Each ring blocked the light from the star twice, once before and once after the star was hidden by the planet. Although initially only five dimmings were seen, more careful examination of the observations showed that a total of nine rings surrounded the planet. They were very close to the planet. The outermost is known as the Epsilon Ring, and its diameter is only about twice that of the planet itself. In contrast to the broad rings of Saturn, those of Uranus are really *ringlets*, which are perhaps only 10 kilometers wide. They resemble Saturn's F Ring. Might there be several, still undiscovered, shepherd satellites keeping the rings in place? Like its satellites, Uranus' rings lie in the plane of the planet's equator. During the course of one Uranus-year, we therefore see them in various orientations, as shown in Figure 74.

After the existence of these rings of Uranus had been established, the Italian professor of mechanics at Padua University, Giuseppe Colombo[1] (1920–1984), discovered indications of the rings on the photographs taken by the Stratoscope II project, let by Robert Danielson. These rings had previously gone undetected.

By using refined techniques, it has now become possible to observe the rings of Uranus from the ground. It is not very easy, because the rings are so close to Uranus that their light is swamped by that from the planet. Again one has to choose a wavelength at which Uranus is fairly faint. At a wavelength of 888 nanometers, methane in the atmosphere absorbs most of the sunlight. Photographs confirm the existence of the rings.

When we have to use the most modern techniques to observe the rings of Uranus from the ground, it appears highly unlikely that Herschel could have observed them. The Munich astronomer Felix Schmeidler has shown — now

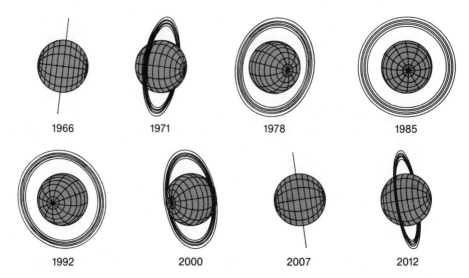

Figure 74. *The aspect of the rings of Uranus as seen from Earth at various times during the Uranus-year. The variations arise for the same reasons as those seen in Saturn's ring (cf. Figure 66).*

that we know the rings' size and orientation in space — that at the time of Herschel's observation, they must indeed have presented just the sort of appearance that he described. According to Schmeidler, it is not surprising that Herschel later rejected the idea of the existence of rings. By then, the side of the rings would have been toward the earth and they would not have been visible, just as the rings of Saturn cannot be seen from the side. Schmeidler points out the agreement between Herschel's possible observations (and lack of them) and the rings' orientation, but leaves the question open of whether it is purely a coincidence or not.

▪ Color Pictures from Three Billion Kilometers

On 24 January 1986, *Voyager 2* radioed images and measurements from Uranus back to Earth. Even before the probe reached Uranus, it had discovered ten previously unknown satellites. They all lie within the orbit of Miranda, the innermost satellite known until then, and orbit just inside and just outside the Epsilon Ring. They include its shepherd satellites (see Chapter 10). All the newly discovered moons are very tiny bodies. The largest of them has a diameter of 150 kilometers, and the smallest one of only 20

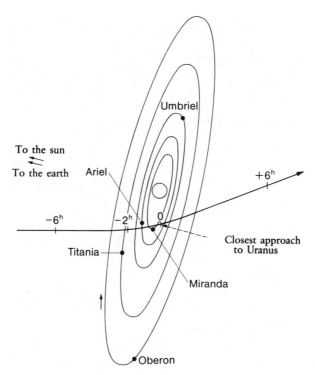

Figure 75. *During its close approach to Uranus on 24 January 1986, Voyager 2 crossed the orbital plane of the satellites nearly at a right angle, approached to within 81,000 kilometers of the planet's upper cloud layer, and a few hours later, passed through the planet's shadow. The figures shown along the path indicate, respectively, the hours before and after closest approach.*

kilometers. The orbit of *Voyager* took it right through the middle of Uranus's satellite system, but kept it outside the rings (Figure 75).

It is not easy to take photographs in the twilight found in the outer solar system. The average exposure time is 10 seconds. The most impressive pictures show the satellites that were already known. Oberon has impact craters that are the centers of bright ray systems. Inside the craters are dark areas, as if some fluid from the interior had flooded the floor of the craters. Titania is similarly saturated with craters, but it is notable for long cracks on its surface. Did the crust of Titania break up at some time in the past when the water in its interior froze — like a bottle of water left out in the cold? Like Titania, Ariel also has long rift valleys running right across some areas of the moon (Figure 76).

Impact craters occur on solid bodies if they are exposed to cosmic bombardment by meteorites for long enough. Cracks and valleys, on the other

Figure 76. *Uranus's satellite Ariel, photographed by Voyager 2 on 24 January 1986 from a distance of 130,000 kilometers. Its diameter is 1200 kilometers. In addition to the numerous impact craters, there are wide valleys and rifts that were created at some time in the past. A few features resemble the "dried streambeds" on Mars (Photo: NASA/JPL).*

hand, show that the interior is not inert, but that it is geologically active. It had been expected that the satellites of Uranus would have been frozen solid ages ago. They were not expected to show volcanoes or signs of the movement of plates forming their crusts. But it was not just Titania and Ariel that contradicted this view. Miranda also exhibited young folds that stretched over broad areas of the satellite's surface, as well as impact craters.

Voyager saw the nine, expected rings of Uranus and discovered a faint tenth, which was not suspected previously. The Epsilon Ring is the most prominent. It is about 30 kilometers wide. The other rings are narrower. The Gamma Ring is only about 600 meters across. Figure 77 gives an idea of how narrow the rings of Uranus are, with the camera looking past the rings at the dark background sky. The rings consist of individual fragments of dust and ice. This was confirmed when the probe went behind the rings as seen from Earth, after closest approach (Figure 74). As the probe passed behind the

Figure 77. *The nine rings around Uranus, photographed by* Voyager 2. *The brightest is the Epsilon Ring, visible in the top left-hand corner. The tenth ring, discovered by* Voyager 2 *lies inside the Epsilon Ring, but outside all the others. It is so faint that it cannot be seen in this image (Photo: NASA/JPL).*

Epsilon Ring, the radio signal was lost for 3 seconds. This indicated that the particles in the ring must be larger than the wavelength of the signal. If the rings merely consisted of dust, the radio waves would not have been absorbed.

But there is dust in Uranus's rings. After passing through the plane of the rings, and while the probe was, so to speak, looking at the rings from "behind," it was looking through the plane of the rings towards the sun. As it passed into the shadow of Uranus, the cameras showed a set of illuminated rings, which consisted of hundreds of individual rings, resembling Saturn's rings. They were rings of dust. In the same way as one can most easily see dust on a windowpane if one is inside the room and the dust is illuminated by the Sun, the normally invisible dust rings could be detected.

On the planet itself, apart from a few clouds, there were no features to be seen. Uranus appears to rotate around its axis in 16.9 hours, and the axis of rotation was confirmed to be as shown in Figure 73.

Long before *Voyager 2* reached the planet, its instruments detected a magnetic field. It was found that the magnetic poles of Uranus are a long way from the rotational poles. They are closer to the planet's equator than to its poles.

As the probe flew past Uranus at a speed of 21 kilometers per second, everything happened very quickly. The encounter lasted only a few hours. Whereas at Jupiter and Saturn the probes were able to fly past the planets and satellites one after the other, at Uranus it was passing through the plane of the system at an acute angle. Thus, it was only close to the planet and the satellites for a few hours. No action could be taken from Earth during the encounter, because any signal from the space probe took 2 hours and 49 minutes to reach us. The images of Uranus that appeared on the screens had been taken long ago. There was no possibility of intervening.

Luckily it had been possible to correct the probe's path before the encounter by a short thrust from the rocket motors. The success of this maneuver was all the greater, because obviously the probe was invisible in even the largest telescopes, and therefore its position could only be determined from its radio signals. The *Voyager 2* mission at Uranus was a complete success. No one expected the satellites of Uranus to have surfaces with such obvious signs of activity. The probe, launched nine years earlier, and described by one American planetary researcher as an aging robot, had again produced good results. *Voyager 2* was now on its way to Neptune, which it reached on 25 August 1989.

▪ Unreliable Uranus

Before a space probe visited it, the most interesting thing about Neptune was the history of its discovery. It all began with Uranus. After Herschel's discovery of the planet, it was not easy to calculate its orbit. It is necessary to observe a celestial body over a considerable portion of its orbit in order to be able to tell whether it is moving around the sun in an ellipse or in a parabola, and also to determine the inclination of the orbit to that of the earth. The slower a body moves, the longer it has to be observed before its true orbit can be determined.

If as bright an object as Uranus were to be discovered today, a search would be instigated in photographic archives of observatories, where more than half a century's photographic plates of all regions of the sky are stored. It would be possible to trace its position on the sky back for decades and to determine its true orbit in an instant. But Uranus was discovered long before the development of photography. The only documentation then consisted of the records that had been made by earlier observers and which had been gathered in certain stellar catalogs. The astronomers had recorded what they had seen through the telescope. But a star that is visible to the naked eye under favorable conditions would not have been missed, even before the invention of photography. The new star had indeed been seen by others before Herschel, but it had been taken for a fixed star. We now know of more than 20 observations of Uranus before Herschel's time.

The Göttingen astronomer Tobias Mayer (1723–1762) had seen Uranus on 26 September 1756 and thought that it was a fixed star in Aquarius. That was 25 years before Herschel. Apart from Mayer, Uranus was also observed on 23 December 1690 by the British astronomer John Flamsteed (1646–1719), who thought it was a fixed star in the constellation of Taurus. These observations, made before its actual discovery, enabled a more accurate determination of the orbit of Uranus. They confirmed that it was a true planet far outside the orbit of Saturn.

But the more observations were accumulated, allowing the calculations of Uranus's orbit to be improved, the worse the planet's motion agreed with the earlier observations by Flamsteed and Mayer. The situation was clearly described by Alexis Bouvard (1767–1843), who published tables of the motions of Jupiter, Saturn, and Uranus in Paris in 1821.

> The preparation of tables for the planet Uranus presents us with the following alternatives: If we combine the ancient observations with the modern ones, the former will be adequately represented, but the latter and their inherent accuracy will not. But if we discount the ancient observations, and retain only the modern ones, we shall obtain tables that will have the desired accuracy relative to modern observations, but which will not conveniently satisfy the ancient observations. We have to decide between these two choices.

For his tables, Bouvard had chosen the latter alternative. This did not meet with the approval of his colleagues, because obviously he regarded the observations of the earlier astronomers as being less accurate.

But after a few years, it became obvious that Bouvard's tables did not agree at all well with the new observations that were being made every year. Was it possible that Newton's law of gravity did not apply over such enormous distances as those between the sun and Uranus? It appeared to account for all the other planets very well, including the asteroids that had meanwhile been discovered. Only Mercury continued to cause a problem (see Figure 41). Had Uranus encountered a comet and been knocked out of its proper orbit? Were there unknown planets outside the orbit of Uranus that were perturbing it?

An important figure in the drama that was about to unfold was Sir George Biddell Airy (1801–1892). In 1834, the man who was later to become the most influential astronomer in England, was professor of astronomy at Cambridge. The Reverend Dr. Thomas J. Hussey (1792–c. 1868), an English amateur astronomer, wrote to him on 17 November 1834: "The apparently inexplicable discrepancies between the 'ancient' and the 'modern' observations suggested to me the possibility of some disturbing body beyond Uranus, not taken into account because unknown." Airy did not take the amateur's letter at all seriously.

In the spring of 1835, Halley's Comet returned. But it did not really behave as expected. In attempting to explain the discrepancy, the idea that an unknown body was causing the perturbation was again raised. On 28 February 1840, the great Wilhelm Bessel (1784–1846) gave a lecture in Königsberg in which he came out against rejecting the ancient observations as inaccurate and thus artificially simplifying the problem of the motion of Uranus. He championed the view that only a new planet beyond the orbit of Uranus could solve the problem. More and more astronomers took this view in the first half of the nineteenth century. A few were sceptical, however, including Airy, who had by now become Astronomer Royal—a sort of astronomical pope. In 1842, the Göttingen Academy of Sciences offered a prize for the solution of the Uranus problem.

▪ The Predicted Planet

The two main protagonists were the Briton John Couch Adams (1819–1892) and the Frenchman Urbain Joseph Le Verrier (see Chapter 5). As a 22-year-old student at Cambridge, Adams had already learned from Airy of the problem concerning the inexplicable motion of Uranus, and of the question of whether an as yet unknown planet might be the cause. Immediately after he graduated in 1843, he devoted himself to the Uranus problem. Airy sent him his latest observations from Greenwich. By September 1845, Adams had calculated what orbit and mass the unknown planet must have to explain all the irregularities in Uranus' motion. Adams wanted to deliver his result to

Airy personally, but as a result of a series of misunderstandings, he was not able to speak to Airy on either of the two trips he made to Greenwich. On his second visit, he left a written description of his results and went away again, disappointed. Airy replied, thanked him for the details, praised his work and posed a counterquestion. By now, Adams was resentful and took a year to answer. During this time, Airy took no action. A whole year went by without anyone searching for the planet that Adams had predicted.

Frequently, scientific problems lie untouched for decades and are then simultaneously taken up quite independently by more than one person. On 10 November 1845, only a few days after Adams had finally replied to England's leading astronomer, Le Verrier presented a paper to the Académie des Sciences in Paris on the motion of Uranus. Le Verrier was originally a chemist and had a good government post in Paris. But when, in 1836, he was about to be moved to the provinces, he took a post as teacher in Paris. As a talented mathematician, however, a year later he was offered a position as lecturer in astronomy at the École Polytechnique. This started his career in celestial mechanics. Airy's counterpart in France was François Arago (1786–1853). He first called Le Verrier's attention to Mercury. This began the work on Mercury that later sent Le Verrier on the fruitless search for Vulcan (see Chapter 5).

In the summer of 1845, Le Verrier turned his attention to Uranus. It did not take him long to calculate the properties of a hypothetical outer planet, the gravitational effects of which would perturb the motion of Uranus. In England, Adams was dogged by bad luck. When he finally answered Airy in September 1846, the latter did not receive the letter immediately, because he had gone to take the cure in Wiesbaden. Adams also missed by one day a meeting of an English scientific society in Southampton, at which he wanted to present his results about Uranus. Now, however, observers in England began, slowly and ponderously, to search for the new planet. The work went ahead very sluggishly.

In France, Le Verrier did not meet with immediate success, however. He did not succeed in prompting his French colleagues to search for the new planet. On 18 September 1846, he wrote about it to Johann Gottfried Galle (1812–1910) in Berlin. "At present, I am looking for a persistent observer," he says in the letter, "who is prepared to sacrifice some time to examining an area of the sky where there is possibly a new planet to be discovered. I came to this conclusion from our theory for Uranus. . . . It is impossible to account properly for observations of Uranus, unless the effect of a new, previously unknown, planet is introduced." In his letter, Le Verrier included the position in the sky.

Galle received the letter on 23 September 1846. It was not so simple for him to fulfill Le Verrier's wish. First he needed permission from his chief; such is the inviolable custom at observatories — or at least it was then.

Johann Franz Encke gave Galle the green light for the search. A younger colleague advised him to use the sheet from a new atlas, the *Berliner Akademische Sternkarten* by Carl Bremiker (1804–1877), that showed the area in question. Although this sheet of the atlas had just been completed, it had not yet been published, but the Berlin Observatory had some proof sheets for correction. That same night, Galle looked at the area of sky through the telescope, but at first could see nothing unusual. Then he took Bremiker's chart, and while he described the field of view, star by star, an assistant compared his observations with the stars on the chart. It was not long before he saw a star that was not plotted on the chart. The next night he checked the object again and immediately established that its position relative to the fixed stars was altered. The predicted planet had been found!

We should not underestimate Bremiker's contribution to the discovery of the planet. In a history of astronomy that appeared in London in 1852, the author Robert Grant wrote about the advantage Galle had with Bremiker's star charts: "If a similar facility of search had been accessible at the Observatory of Cambridge, it is admitted by all persons that the planet could not fail to have been first discovered there."

Who did really predict the new planet? The position in which it was found by Galle was close to the spot suggested by Le Verrier, but the position predicted by Adams was also near by. Adams had obtained his result a year before Le Verrier, but had only informed Airy and had not published it. For a time, there were angry waves of patriotism in England and France. But Adams, who, through no fault of his own, had lost, showed British fairness. In a book published later, he described exactly what stage his work had reached at specific times. "I mention these dates," he wrote, "merely to show that my results were arrived at independently, and previously to the publication of those of M. Le Verrier, and not with the intention of interfering with his just claims to the honours of the discovery; for there is no doubt that his researches were first published to the world, and led to the actual discovery of the planet by Dr Galle." Adams and Le Verrier met each other for the first time in 1847, and struck up a friendship that was to last until Le Verrier's death in 1877. After his successful prediction of Neptune, Le Verrier returned to the problem of the motion of Mercury, which sent him on a wild-goose chase after the nonexistent planet Vulcan, which we discussed in Chapter 5.

The new planet was given the name Neptune. It could be seen with small telescopes (the diameter of Galle's telescope was only 25 centimeters). As early as 10 October 1846, William Lassell—we already met him as the discoverer of two of the satellites of Uranus—found its satellite Triton, which orbits Neptune in about 5.9 earth-days. It was only in 1949 that Gerard Kuiper found the second satellite of Neptune, Nereid, which takes 360 days to orbit the planet. But Lassell not only saw Triton, he also thought he had seen a ring around Neptune. Other experienced observers also felt they had seen such a ring.

▪ *Did Galileo See Neptune?*

After Neptune's orbit was determined, naturally people checked to see if any astronomer before Galle had already seen the planet. It turned out that it had been seen three times from England in the search for Adams' planet, but had been taken for a fixed star. As, unlike Galle, no one was lucky enough to have had a good star chart, no one had thought any more about it.

Quite recently it has been suggested that a famous observer might have seen it a long time ago. When the astronomer Charles Kowal and the historian of science Stillman Drake checked earlier positions of Neptune, they noticed that Neptune and Jupiter must have been very close to one another in the sky in 1613, and therefore, any observer looking at Jupiter must have had Neptune in the same field of view. In 1613, there were few telescopic observers, but one was Galileo. Kowal and Drake actually found drawings by Galileo of the positions of the satellites of Jupiter for 28 December 1612 and 28 January 1613. The drawings, made at the telescope, showed Jupiter and three of the satellites at any one time; but they also showed a star, which Galileo took to be a fixed star. It was probably Neptune.

▪ *Neptune and Its Ring*

Neptune's distance from the sun is thirty times that of the earth. It takes the planet nearly 165 years to complete one orbit, and so it has not quite completed one orbit since its discovery. Its mass exceeds that of Uranus, with about 17 earth-masses in a body 49,000 kilometers in diameter, and its density of 1.65 grams per cubic centimeter is higher than that of Uranus. Neptune's atmosphere appears to be similar, containing hydrogen, helium, ammonia, and methane.

Neptune's day might well be described as Neptune's night, because its great distance from the sun means that the brightness of the sun is about one one-thousandth of what it is on Earth. But it is still about five hundred times brighter than full moon. Despite the low solar radiation, the atmosphere appears to be fairly active, and it is suspected that Neptune, like Jupiter and Saturn, may heat its atmosphere from beneath.

Neptune's ring suspected by Lassell, the amateur astronomer, did not survive very long. The Director of the Cambridge Observatory did think he saw it two years later, but it was never seen again, despite searches with increasingly better telescopes.

On 24 May 1981, Neptune occulted a fixed star. It was noticed that the light from the star became distinctly fainter 8.1 seconds before the star vanished behind Neptune. It was concluded that this indicated the existence of a third satellite of Neptune, which orbited the planet at a distance of three Neptune radii.

On the night of 21 to 22 July 1984, the Munich astronomer Reinhold Häfner and the Belgian Jean Manfroid were observing Neptune from the European Southern Observatory at La Silla, north of Santiago in Chile. That night, Neptune was due to pass close to a star, which was a very cool object and particularly bright in the infrared region. The star appeared brighter in the infrared than Neptune. The two astronomers were observing with different telescopes and were recording the combined light of Neptune and the star. At 5:40 Universal Time, it became 35 percent fainter for 1.2 seconds. Both observers, using different telescopes in different domes saw the event on their chart recorders. As Neptune could not possibly show variations in its own light with a period of fractions of a second, it must have been the light from the star that was blocked for a short time.

Months after the European observers had published their results, American observers found the same event on magnetic tapes, on which they had recorded Neptune's apparent close approach to the star on the same night. Their observations were made only 100 kilometers south of La Silla, at the Inter-American Observatory at Cerro Tololo.

Examination of all the results, including those of 1981, suggested the existence of a ring, about 10 to 15 kilometers wide, surrounding Neptune at a distance of three planetary radii. The ring did not appear to be very uniform, because it should have occulted the star twice, once before and once after Neptune itself occulted the star, and the light should have dropped twice. But that did not happen. If the ring is imagined to consist of lumps of ice orbiting in a loose formation, on one occasion the ring of ice must have cast its shadow on the earth when blocking the star. During the second passage in front of the star, the blocks of ice must have been so widely separated that none of them intercepted the line of sight between the earth and the star. When Neptune again occulted a faint star on 23 April 1986, no eclipses of the star by the ring were detected either before or after the main event. Not much more was found out about the ring until *Voyager 2* visited the planet in August 1989.

▪ Voyager 2 *at Neptune*

There was great excitement at the Jet Propulsion Laboratory in Pasadena as the day arrived on which *Voyager 2* was to encounter Neptune, its last target, after its 12-year journey. It was not easy to follow events that were occurring

4.4 billion kilometers away. The faint signals took four hours to reach the earth and could only be detected with the largest radio antennas.

Anyone who thought that Neptune was a cold, frozen world, must have been surprised at how active it was. A gigantic circular storm in Neptune's atmosphere gave rise to a dark spot, the Great Dark Spot, which was the size of the earth. It closely resembled its red counterpart on Jupiter (Color Plate 7). White shreds of cloud surrounded it, and others raced at speeds of hundreds of kilometers per hour along the parallel pattern of belts that — as on Jupiter — stretched around the planet. The planet rotates around its axis in just under 16 hours. As with Jupiter, the energy that drives the activity in its atmosphere appears to arise within the planet's interior.

Its magnetic field doubtless also has its source there. As *Voyager 2* approached Neptune, its receivers picked up radio bursts. They appeared to be produced by electrically charged particles trapped in Neptune's magnetic field, just like the particles within the earth's radiation belt. The radio bursts occurred with a 16-hour cycle, reflecting the planet's rotation period.

Voyager 2 not only confirmed the observations made from Earth that suggested that Neptune was surrounded by a ring; in all, eight narrow rings were discovered by the probe, together with six previously unknown satellites of the planet.

Five hours after its closest approach to Neptune, the probe flew at a distance of 38,500 kilometers past Triton, the satellite discovered nearly 150 years before by William Lassell. If people were surprised by Neptune, Triton proved to be even more amazing. "Triton steals the show," wrote the British science magazine *New Scientist* a week later. The reason for this was the ice volcanoes that were detected on the surface. Triton's diameter of 2720 kilometers is slightly less than that of the moon. In contrast to the dark face of the earth's satellite, the snow-covered surface of Triton reflects well over half the sunlight falling on it. Only a few impact craters can be detected on the images returned, but furrows and ridges run across the surface for hundreds of kilometers. There are also plains covering hundreds of square kilometers that resemble sheets of solidified lava (Color Plate 8).

Triton's atmosphere consists primarily of nitrogen, but some methane appears to be present as well. A haze layer reaches heights of some 14 kilometers above the surface. It seems that Triton's surface was fluid at some time in the relatively recent past. The people who suggested that perhaps Triton was once an independent planet, captured into an orbit around Neptune by the combined gravitational effects of the planet and of the sun, are probably correct. Strong tidal forces would have kneaded the erstwhile planet when it became a satellite, and would have heated its interior until its surface melted. Craters left by earlier meteoritic impacts would have been obliterated. Later, after its orbit had evolved toward a circle, the tidal forces would have diminished, and the features observed today would have been created. It is possible that beneath the crust, nitrogen is still liquid today, and that occasionally "volcanoes" erupt liquid nitrogen to flood across the icy landscape.

▪ *A Young Man from Kansas*

On the afternoon of 18 February 1930, the 24-year-old Clyde William Tombaugh looked at the plates taken on 23 and 29 January. They were not in sequence, but since they were better than the others, he put them in the blink-comparator. This piece of equipment is used to compare two photographic plates of the sky with one another, in order to see whether anything has changed in the interval between the two exposures. He had taken the plates himself with a 33-centimeter-diameter astrograph—a large astronomical camera.

Clyde Tombaugh had been at the observatory in Flagstaff, Arizona, for rather more than a year. He liked the work, and it seemed as if he would keep his job for some time yet. He had been rather uneasy on 25 October the previous year. That was "Black Friday," when the New York Stock Exchange collapsed. But he was relieved to learn that the money that funded the observatory and its staff was securely invested.

In recent years, Clyde had lived at Burdette in Kansas and after leaving high school there in 1925, he had worked on his parents' farm during the summer. In the winter, he found time for his hobby, building astronomical telescopes. In the spring of 1928, he applied for a post at the Flagstaff Observatory, enclosing some drawings of Jupiter that he had made through his telescope. The observatory was about to hire someone to take plates of the sky, so in February 1929 he took up a post at Flagstaff.

When the young Tombaugh had put the two plates, taken at an interval of six days, into the comparator and, by using a lever, flicked first one and then the other into the field of view, he immediately saw a point of light jumping about. On the negative plate, the stars appeared as black dots, the brighter ones having larger diameters. This star was a tiny point that had moved slightly between the two exposures—not much, only about 3.5 millimeters. By switching from one plate to the other, the point jumped back and forth. Clyde Tombaugh had long since learned to recognize plate blemishes, and he also knew that over an interval of six days, asteroids would normally move a much greater distance. What he saw was moving too slowly to be an asteroid between Mars and Jupiter. The object must be orbiting much farther out in space. It must be Planet X, which was the reason for his being appointed. Tombaugh informed his superiors. But they first wanted the object to be confirmed by other observations. That evening the sky was cloudy, and instead of taking other plates, young Tombaugh went to see a movie in Flagstaff. He recalled later that they were showing a Gary Cooper film, *The Man from Virginia*. The next evening, however, the object was again photographed. The long-sought Planet X had been found.

▪ *The Search for Planet X*

This ended a search that had begun nearly 30 years before and which is closely linked with the name of one man, who had passionately and impatiently used every possible means to search for the new planet, but who did not live to see his efforts crowned with success.

We have already met Percival Lowell in connection with his search for the Martian canals. Equally hectically he carried out a search for a planet outside Neptune's orbit. He did this by employing two parallel methods. In one, he tried to predict the position of the planet from the irregularities in the motions of Uranus and Neptune, just as Adams and LeVerrier before him had done in the case of the irregularities in the motion of Uranus. He did not just work persistently on his own, but engaged numerous assistants, who carried out the calculations for him. At the same time, he had the observatory he had built at Flagstaff in Arizona keep a photographic watch on the sky, in the hope of detecting an object, moving like a planet at the edge of the solar system. In 1915, he ceased his theoretical work on Planet X, as he called the unknown planet. He did not suspect that he already possessed two plates that showed it as a very faint object. He never knew, because they were found only 14 years after his death when, in 1930, the orbit of the newly discovered planet was calculated backward in time, and a search was made on old Flagstaff plates.

Planet X, found fairly close to the position that Lowell had predicted, now needed a name. The decision lay with the director of Flagstaff Observatory, Vesto M. Slipher (1875–1969). Suggestions poured in from all over the world. Zeus, Odin, Cosmos, and Lowell were just a few of them. On March 14, a Mr. F. Madan wrote to the professor of astronomy at Oxford that his granddaughter Venetia Burney had suggested the name Pluto. Slipher decided to adopt the suggested name, not least because the first two letters were Percival Lowell's initials.

The astrologers, however, paid no heed to Percival Lowell's claim to be recognized in the name. They preferred to think of Pluto as the god of the Underworld. One of them even linked the name with the atomic bomb—because of plutonium.

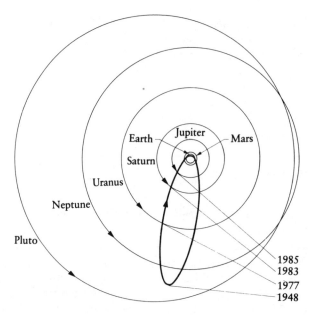

Figure 78. *The orbits of the outer planets and of Halley's Comet. The figures shown on the comet's elongated orbit give the positions of the comet at various times during its last return. The diagram also shows that, over a certain portion of its orbit, Pluto is nearer to the sun than Neptune.*

▪ *Pluto*

Even today, more than half a century after its discovery, Pluto has only covered a small fraction of its orbit.[2] It was discovered in Gemini and is now (in 1989) in Libra, close to the borders of Virgo and Serpens Caput.

Astronomers can determine Pluto's orbit fairly accurately from the small portion that it has traversed since 1930. This shows that Pluto is somewhat of an exception. Its orbit is so strongly elliptical that for 20 years of its 248-year period, it is closer to the sun than Neptune. It is therefore not always the most distant planet in the solar system (Figure 78). In fact, since 1979, Neptune has been the most distant. Pluto will not move outside the orbit of Neptune again until 1999.

But Pluto is also an exception in other respects. Whereas the orbits of all the other major planets lie more or less in the same plane, the plane of Pluto's orbit is inclined at an angle of 17°. Therefore, Neptune and Pluto can never collide, since they never approach to within less than 18 times the Earth–Sun distance. Pluto's remarkable orbit did suggest that Pluto might be an escaped satellite of Neptune. It may once have orbited Neptune together with Triton and may have been flung out into space, just as the space probe *ISEE-3* was ejected from its orbit near the earth (see Chapter 2). Because of its low mass, Pluto may be compared with the Galilean satellites of Jupiter rather than with a true planet. In 1978, James Christy discovered Pluto's satellite Charon on plates taken that year at Flagstaff Observatory. The moon orbits Pluto at a distance of 20,000 kilometers in 6 days and 9.3 hours and can be seen on the photographs only with some difficulty. The popular-science author Michael Hahn has written; "The difficulty of the observation may be appreciated if one imagines that one has to recognize the headlights of an automobile that is 200 kilometers away as being separate, and when one of them is only showing a parking light."

Despite this difficulty, the orbital motion of the satellite enabled Pluto's mass to be determined. Pluto has only a few thousandths of the mass of the earth. This is so insignificant that Pluto cannot be responsible for the irregularities in the motions of Uranus and Neptune, and therefore its discovery at the position predicted by Lowell must be a coincidence. It cannot be likened to the discovery of Neptune. Since then, Neptune has moved farther along its orbit, so its motion is better known and the observational accuracy has improved. Neptune and Uranus appear to move in a manner that is in accordance with the combined gravitational fields of the sun and the planets. Some of the irregularities that were earlier thought to be present in Neptune's motion arose because of uncertainties in determining its position in the sky, for which it is necessary to use the fixed stars that appear in the background. But in Lowell's time, their positions were not as accurately known as they are today. Even today, although Neptune may not appear in exactly the predicted position, the irregularities lie not so much with Neptune, but with the positions of the fixed stars contained in our catalogs.

Are there still other planets orbiting the sun beyond the orbits of Neptune and Pluto? How can we find out anything about these objects, which are probably so far away that their gravity would have hardly any effect on the known planets? They are undoubtedly so faint that they can only be photographed with the largest telescopes. Despite this, after his success with Pluto, Tombaugh continued to search for transneptunian planets. In a long series of investigations until the Second World War, he spent a total of about 7000 hours comparing plates in the blink-comparator. He spent additional thousands of hours at the telescope, taking photographs, with the exposure time for each plate between 60 and 90 minutes. But his search was without success.

The comets might offer a way of gaining some knowledge about space outside the system of known planets. Many of them move on orbits around the sun that are more elongated than that of Halley's Comet (Figure 78). For example, Comet Mrkos (shown in Figure 53), takes 353 years to complete one orbit. The point at which it is farthest from the sun lies at about twice Pluto's distance. If we assume that all comets are derived from the Oort cloud (see Chapter 7) and that their orbits have been turned into ellipses by the effect of the planets, then elliptical cometary orbits that reach as far out into space as that of Comet Mrkos may somehow be related to more distant planets. The longer the period of a comet, the farther its orbit stretches out into space. This follows from Kepler's third law (see Chapter 1). So, details of the orbits of long-period comets have been tabulated and searched to see if hypothetical transplutonian planets might be responsible for the fact that these comets now move in elliptical orbits. One astronomer has even given the transplutonian planets romantic names, such as Hades, Persephons, and Teiresias. There is only one problem: No one has seen any of them.

TWELVE

The Nebula from Which the Sun Was Formed

I assume that all the material of which the spheres that belong to our solar world, all the planets and comets, now consist, was, at the beginning of all things, dispersed into its elementary components, which filled the whole space of the world structure, within which these accumulated bodies now revolve.

IMMANUEL KANT

■

The family of planets with the all-dominant sun at its center prompts us to ask: How did it all come into existence? Did the sun give birth to the planets? Did all of them arise from the sun? Nature does seem to give us some indications of the answers to these questions.

The planets and the sun exhibit certain regularities that cannot be accidental. They must be related to the formation of the system as a whole. Anyone who wants to discover how the planets were formed will find that certain clues have been left behind. At present, we are not very good at interpreting these clues, and therefore we only have a very approximate idea of the events that occurred in that remote past when the earth did not exist, when no other planets and satellites moved around the sun, and when the sun itself probably did not exist.

▪ *The Clues*

Most of the mass of the system formed by the sun and the planets is to be found in the sun. If one were to collect the material in all the planets into a single body, it would be a very tiny object by comparison. It would take 750 such bodies to obtain the mass of the sun.

The elliptical orbits of the planets exhibit a form of order. Their orbital planes are not precisely the same, but their inclinations relative to one another are comparatively small (Figure 79 *bottom*). The greatest deviation is shown by the orbit of Pluto, but the outer bodies in any system always appear to be exceptions to a greater or lesser degree, as we can see from the outer satellites of the planets. The innermost planet, Mercury, also disobeys the general rule to a certain extent.

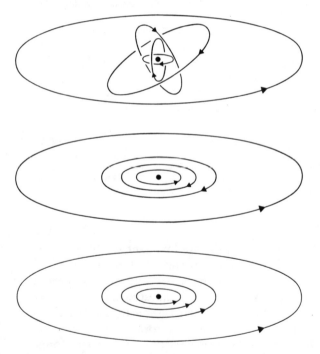

Figure 79. *Conceivable planetary orbits around the sun.* Top: *The elliptical orbits are inclined at various angles to one another.* Center: *The orbits all lie in approximately the same plane, but the planets themselves do not all move around the sun in the same direction.* Bottom: *The orbits are nearly all in the same plane, and the motions are all in the same direction.*

It is important to note that the laws obeyed by the motions of the planets are of two completely different types. The fact that the planets move in ellipses around the sun, in accordance with Kepler's second law, is one type of order, which indeed allowed us to determine the law of gravitation but which tells us nothing about the origin of the system. No matter how the planets were formed, they would have to obey Kepler's laws. These laws would still be obeyed if the planets moved around the sun in the way shown in the top part of Figure 79. The fact that the solar system actually behaves as shown in the bottom part of Figure 79 tells us that when the solar system was formed, nature chose one specific possibility from among many. Yet another special choice occurred in that the planetary orbits could have been any more or less elongated forms of ellipse. But they are not; the orbits of the major planets are almost all very close to circular. Only Pluto and Mercury are exceptions.

It would also be perfectly permissible under the laws of mechanics — and quite in accordance with natural laws — for some of the planets to circle the sun in one direction, but for others to do the opposite, like two-way traffic (Figure 79 *center*). But the planets, even erratic Mercury and troublesome Pluto, all revolve in the same direction, like cars on a racetrack. They have even more in common: The direction in which they orbit is the same as the sun's rotation (which is once in about 27 days) and the rotation of nearly all the major planets, only Venus and Uranus being the exceptions.

The satellites also generally orbit in the same direction around their planets. It seems that planets with many satellites, particularly Jupiter, Saturn, and Uranus, form mini-solar-systems in themselves, with most of the satellites orbiting in nearly the same plane and in the same direction around their individual central bodies, which generally rotate in the same direction.

The largest planets rotate fastest, and the smaller ones more slowly. The original rotation of Mercury and Venus has obviously been modified by tidal friction. They now rotate in a completely different manner from the way they did when they were formed.

The chemical compositions of the inner planets are different from those of the outer ones. The planets as far out as Mars consist of material similar to that found in the earth. Their densities do not differ much from that of our planet. Elements like silicon, nitrogen, oxygen, and iron play important parts, but not hydrogen and helium. That is certainly not generally true throughout the universe. The sun, for example, consists primarily of hydrogen and helium and shows only traces of the materials that form the inner planets. The outer planets are also completely different. The abundances of the elements found in Jupiter, for example, are probably the same as those in the sun.

The chemical properties of the planets give some hints as to their origin. One fundamental property of the solar system, however, remains the way in which the planets rotate around the sun.

▪ *The Division of Mass and Angular Rotation*

Within the system consisting of the sun and the planets, each body has angular momentum. In Chapter 3, we spoke about the angular momentum of an ice skater doing a pirouette. As in the skater's case, in order to find the angular momentum of the sun, we need to calculate the rotational velocity times the distance from the axis for every atom, and then obtain the overall sum. But consider a planet, such as Jupiter. It has two forms of angular momentum. First, it revolves around the sun. This means it has *orbital angular momentum*, which we obtain if, for each of its atoms, we calculate the value of the velocity times the distance from the sun and take the overall sum. Second, it rotates around its own axis. If we take the rotational velocity times the distance from its axis of rotation for every one of its atoms and sum the results, we obtain the planet's *rotational angular momentum*. We mentioned already that we do not actually have to carry out this calculation for each atom in order to obtain the rotational angular momentum.

If we take the sum of the orbital angular momentum of all the planets, the total orbital angular momentum thus obtained is much greater than the rotational angular momentum of the sun. When the solar system was formed, nearly all the angular momentum went into the planets. The sun had only a very small fraction, in fact, about 2 percent. But, as we have already seen, the sun contains nearly all the mass. However the solar system was formed, the sun gained the mass, and the planets gained the angular momentum.

We have already seen in our earlier discussion of the specific properties of the planetary system that rotation played an important role when the planets were formed. This was noted by René Descartes, who therefore suggested that the sun and the planets were formed from a flat, rotating disk of gas. We shall see that in our days of fast computers and interplanetary exploration, we are not much farther advanced than the early French philosopher.

▪ *Immanuel Kant's Self-Contained Cloud*

On 14 March 1755, the 31-year old Immanuel Kant dedicated his *Universal Natural History and Theory of the Heavens* to the king of Prussia. In this book, he tried to explain the "whole world picture," using Newtonian mechanics. His work gave the first hints of how things may have occurred.

Kant begins with a simple initial state for the universe and tries to determine what would be formed from it with the passage of time. "After assuming that the Universe was initially in the simplest form of chaos," he wrote, "I have employed no other forces than gravity and the force of repulsion to develop the great order that exists in nature. . . ."

According to him, there was originally a "primordial nebula," a disorganized cloud of fine dust particles and bodies similar to our meteorites. The spherical cloud was held together by its own gravity. Anything within it had no orderly motion, and the individual particles were moving in all directions. Only when a particle was on the point of leaving the cloud would it be restrained by the cloud's gravity.

As the particles were flying around in completely random orbits, like a gigantic swarm of gnats, they repeatedly collided with one another. When two particles, moving in opposite directions, crash into one another, they lose their velocity and tend to sink toward the center. Thus, material continuously streamed toward the central regions of the cloud, where it would have formed a massive central body, the sun. In the outer regions, the particles collected into condensations, which then orbited the sun. They attracted other particles flying around in the cloud and swept them up. Over the course of time, the space originally filled by the cloud would have been cleared by the planets as they accumulated.

Kant's scheme for the formation of the solar system does not explain the most important of its properties. The planets do not have to move around the sun in the same direction, and their paths do not have to lie in nearly one plane.

▪ Pierre Simon Laplace's Rotating Disk

The French mathematician Laplace envisaged the solar system as beginning with a slowly rotating, spherical cloud, which stretched far beyond the present planetary system. It was a gas cloud, but the motions within it were not chaotic. Everything rotated slowly and uniformly around its axis. But the cloud's gravity tried to pull all the atoms of gas toward the center, and the cloud began to shrink. In order to understand what follows, we have to remember the principle of the conservation of angular momentum, described in Chapter 3. As the sphere shrank, its atoms moved nearer to its axis of rotation. Because the product of distance from the axis times rotational velocity cannot change, the rotation of the cloud had to become faster. Like the ice skater, who pulls in her arms, the cloud rotated faster and faster as it contracted. But because its rotation became faster when it shrank, centrifugal force eventually became significant. It was strongest at the equator of the rotating cloud and acted outward against the contraction. Thus, the cloud mainly contracted along its rotational axis and became flatter. After a period of time, the sphere took on the form of a pancake (Figure 80).

From then on, everything occurred within a disk. As the diameter shrank further, the stage was soon reached when the outwardly directed centrifugal force balanced the inwardly directed force of gravity. The outer layers of the cloud could no longer shrink. As a result, a ring of material was left behind in the plane of the disk, whose remainder continued to shrink. Soon centrifugal force again became too great, and the next ring was left behind. The planets then formed from these rings. In this scheme, the outer planets are the oldest, and the inner ones were born later.

In Laplace's scheme of the formation of the solar system, it is self-evident that the planetary orbits would all be in nearly the same plane, and that the planets would orbit the sun in the same direction, namely in the direction in which the cloud originally rotated.

The problems with Laplace's theory occur in the details. It does not explain why the disk, after having shed one ring, should have continued to shrink. But there are other problems. If, when the outer planets were formed, the solar disk was already rotating so fast that at the equator its gravity could no longer counteract the centrifugal force, then the sun should still be a flat pancake. In fact, however, it rotates so slowly that its flattening cannot be measured. The rapidly rotating, flat disk from which the sun eventually formed must have been braked somehow. To be more precise: It must somehow have shed its angular momentum. Laplace's theory gives no information about how that occurred. We must therefore turn to modern ideas about the formation of the planets.

▪ By-Products of the Sun

We cannot envisage the planets as having been formed without the sun. It is the body that rules over all the others, and which they all orbit in a single direction. But the sun is just *one* star among one hundred billion that populate our Milky Way system. How did they form? Can we learn anything about the birth of our own sun from the history of their formation?

It took many generations of astronomers to develop the picture of the formation of the stars that is generally accepted today. A lot of observations were required, not just in the visible region; infrared and radio telescopes proved of vital assistance as well. Powerful computers had to carry out the calculations about the evolution of stars, which show what newly formed stars are like. Finally, complicated computer simulations have helped to find the currently accepted solution to the problem of the actual formation of

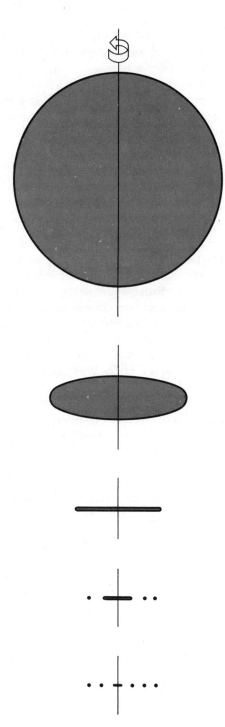

Figure 80. *A diagrammatic represen- tation of the formation of the solar system according to Laplace. A slowly rotating cloud contracts. It therefore rotates faster and flattens out. After a time, it has become a flat disk. If it attempts to contract still further, ma- terial must be left behind in a rotating ring, which will then accumulate into a single body, a planet. Over the course of time, more rings are formed, and thus more planets.*

stars.[1] This was already mentioned in Chapter 9 in connection with Jupiter's internal heat source, and I will briefly summarize the main points of this modern theory here.

Stars arise from clouds of gas and dust, similar to those found between the stars today. When a gas cloud is sufficiently massive and cold, the mutual attraction of its individual components becomes great enough for the cloud to start to collapse. This collapse is driven, and accelerated, by the cloud's own gravitational attraction. After a few hundred thousand years, a hot core forms within the cloud, and within this cloud the internal pressure resists further collapse. This is the embryo from which a star will form. As yet, it still contains only a small fraction of the mass of the cloud, the majority of which is still collapsing. In the course of time, the whole cloud eventually "rains down on" the core, whose central regions have, in the meantime, become so hot that fusion of hydrogen into helium has begun. Nuclear reactions began in the sun approximately 4.6 billion years ago, and all the energy that we receive from it is nuclear energy.

But the process just described is an oversimplification. The material from which the solar system was formed was derived from the Milky Way system. This is a disklike formation of stars, within which gas and dust clouds are observed. It rotates about its center, and every molecule of gas, every grain of dust, possesses angular momentum. If the solar system was formed from gas and dust in the Milky Way, then we cannot begin with a spherical cloud that just collapses. As in Laplace's theory, we must begin with a *rotating* cloud, even if its initial rotation is extremely small. As the cloud collapses, its angular momentum is conserved, and it must rotate faster and faster. Centrifugal force rises as a result, until it eventually compels the collapse in the equatorial plane to come to a halt. Instead of a spherical core in the center, onto which the rest of the cloud collapses, there is a lens-shaped object, a rotating nebula, similar to Laplace's disk. At its center, the sun forms from material that has a relatively low angular momentum, but which is still rotating very rapidly and, therefore, is probably strongly flattened. The material in the disk rotates so rapidly, however, that it cannot collapse onto the sun; planets will form within the disk.

In this picture, the planets form from precisely the same sort of material as the sun. Only its greater angular momentum prevents the planetary material from combining with the sun. At first sight this appears contradictory. We said that the sun chiefly consists of hydrogen, which has been partially converted to helium in its interior. The earth, however, consists primarily of rocky materials, with its interior probably containing a lot of iron. Hydrogen and helium are certainly not the primary components of the earth. The same applies to the moon and Mars, and to Mercury and Venus, whose densities equally indicate that hydrogen is not their main component. How can the sun and the planets have formed from the same sort of material and yet be so different chemically? We shall have to conduct yet another of our thought experiments.

▪ Solar Material in the Laboratory

By using spectral analysis to investigate the light from the sun, astronomers have been able to determine the chemical composition of its outer layers. Let us imagine that we could scoop a kilogram of material out of the sun. This would contain 736 grams of gaseous hydrogen, 250 grams of helium, 7 grams of oxygen, and 3 grams of carbon, as well as 1.5 grams of iron, but only 1 gram of nitrogen. That would not be all, however. In our sample of solar material, we would also find traces of all the known chemical elements. The material appears to be completely different from the material forming the earth, the moon, and the meteorites. But it only appears so at first sight.

The matter in the sun is not some mysterious material. We know the chemical and physical properties of its gases. Let us imagine that some of this material, at a temperature of a few thousand degrees, is sealed inside a container. Let us now try to understand what happens when our solar material slowly cools. In order to learn something about what happened in the nebular disk that surrounded the sun from the processes that occur in our solar material, let us assume that the pressure is as low as in the primordial nebula. That corresponds to about one one-thousandth of the air pressure on the surface of the earth.

Initially, as long as the temperature is above 2000°C, we have a hot gas, which is continually and thoroughly mixed by the rapid motion of the atoms of the gas. If we now lower the temperature to below 1900°C, the first solid particles form within our solar gas. The elements osmium, rhenium, and zirconium condense and form fine dust grains. When the temperature falls below 1700°C, atoms of aluminium combine with oxygen atoms, forming grains of aluminium oxide. The lower the temperature, the more types of grains form. When we get near 0°C, nearly all the chemical elements have condensed into solid grains. Only hydrogen, helium, and a few other volatile materials are still in the form of gas.

If we now removed the solid condensates and allowed the remaining gas to blow away, we would obtain a mixture of materials that is far closer to the material occurring in the earth than the original solar gas. The earth therefore consists of a form of "condensed" solar material. These similarities are even more striking when we examine rocks that probably originated in the primordial nebula.

▪ Condensed Pieces of the Primordial Nebula

Each year, about 10 to 20 meteorites are found on earth. They enable us to investigate material from space in a test tube. Approximately three quarters of all meteorites belong to the class known as *chondrites*. The name comes from the Greek word for "grain." They contain grains of harder material that are embedded in a softer matrix. The chondrites attract attention because of their chemical composition. We find that, unlike the earth's rocks, they contain nearly all the chemical elements in precisely the same abundances as those occurring in solar material. The only exceptions are hydrogen and helium. All the hydrogen that is not bound into other solid material appears to have been lost, just as we blew away the remaining hydrogen at the end of our thought experiment. Since the noble gas helium does not combine with other materials, not even with solid condensates, this element has also been blown away from the chondritic material.

It seems that the chondrites are condensates from the primordial nebula. We must therefore assume that the hydrogen and helium in the primordial material were blown away. But, where did the wind come from when the material in the primordial nebula still consisted mainly of hydrogen?

▪ The Primordial Nebula and the Gale from the Sun

In discussing cometary tails, we have already seen that there is a continuous stream of material coming from the sun. The solar wind would never have sufficed, however, to blow all the hydrogen and the helium away from the primordial nebula. But there are signs that in the past, the sun expelled gas into space at a far higher rate.

We no longer see this in the sun, but many stars are younger than the sun. Among recently formed stars, there is a group whose members were originally noticed because they varied irregularly in brightness. We know of many such *variable stars*. Here we are dealing with a subgroup, known as the *T-Tauri stars*. Detailed investigation of these stars has shown that they are expelling streams of gas at velocities of 200–300 kilometers per second. It seems that such a star blows hundreds of thousands of tons of gas out into space every second. If one of these stars is surrounded by a primordial nebula, which will later form a planetary system, then everything that has not condensed into sufficiently large bodies will be blown away into space. If the sun also went through a T-Tauri phase, it would have blown away all the gases in the surrounding nebular disk, and only the solid materials that were contained in larger bodies already orbiting the sun would have remained. There are other indications that such bodies existed.

▪ *The Evolution of the Primordial Nebula*

Let us assume that the primordial nebula has just formed from the surrounding gas cloud, and that the sun has formed in the center around which the nebular disk is rotating. Hydrogen and helium are still the main components of the nebular disk. The heavier elements have condensed into solid dust particles.

The disk of gas rotates around the central sun, and because each particle has to obey Kepler's third law, the inner portion of the disk rotates faster than the outer. As it does so, dust particles within the gas of the rotating disk repeatedly collide with one another. They tend to clump together, and so the solid dust particles become larger and larger. But, just as a flow of air picks up only the smaller particles of dust, in the primordial nebula only the smaller solid grains followed the motion of the gas. The larger the particles became, the more they obeyed only gravity and centrifugal force. Although the finer dust particles continued to be swirled around in the disk by the

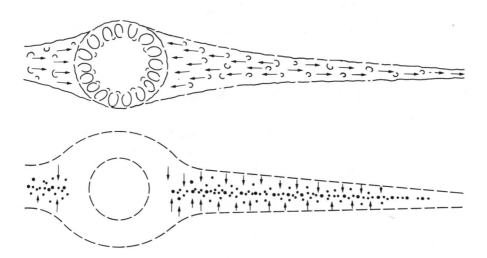

Figure 81. *How the solar system may have been formed.* Top: *Within a slowly rotating cloud, a central body and a flat disk are formed. The disk contains gas and dust like that in the original cloud.* Bottom: *In the center of the disk, the sun has already formed. The dust particles are affected by its gravity but do not fall into it because of the centrifugal force caused by their revolution. Since gravity does not just act in the direction of the rotational axis but also toward the central plane of the disk, this is where the particles collect, clump together, and form larger bodies. The remaining gas is blown away when the sun becomes a T-Tauri star.*

constant movement of the gas, the larger ones settled in the center plane (Figure 81), where they formed a sheet of solid bodies. It is estimated that they were a centimeter and more in size, orbiting the sun like miniature planets.

In this scheme, we have a rotating disk of gas with a rotating sheet of solid, centimeter-sized bodies in its central plane—a sort of gigantic form of Saturn's ring. Mutual attraction between the components of this ring causes the bodies to continue to clump together. It is thought that they may be able to grow to about one kilometer in size. Now they are similar to asteroids. Their gravity captures other passing bodies, and they continue to grow. They thus become true planets. In a few cases, the bodies become so large that their gravity enables them to capture more gas from the disk than their own mass of solid material. This is how the larger planets formed, which mainly consist of gas trapped in this way. As the gas collapsed onto the planets, it heated up, which is why the interiors of Jupiter and Saturn are still warm.

Now the sun becomes a T-Tauri star. Material flows away from it at high velocities, sweeping the solar system clean. Everything that has accumulated in the form of solid bodies remains, as do the masses of gas in the largest planets and in the atmospheres of nearly all the planets. Their gravity enables them to withstand the effects of the strong solar wind.

■ Was the Primordial Nebula Hot or Cold?

In recent years, a nice, simple picture of the processes that occurred in the primordial nebula has been developed. It is connected with the thought experiment that we carried out with solar material, and was put forward by the great cosmo-chemist Harold Urey (1893–1981). Let us think of the primordial nebula as being hot close to the Sun, with temperatures of around 1400°C, but with a temperature of about −100°C out where Uranus and Neptune formed. Then at the distance of Mercury, the only materials that could condense would be those that, in our thought experiment, became solid very early on. Mercury should therefore consist of such materials. Farther out, where the temperature of the primordial nebula was lower, materials with lower condensation temperatures would solidify. This picture might therefore explain the decreasing mean density of the planets with increasing distance from the sun.

Recently, however, some objections have been raised to the theory of the hot primordial nebula. Minerals are found in meteorites that would have been distributed throughout the entire meteoritic body if it ever had a temperature of nearly 2000°C. But they only occur in the form of isolated inclusions. From this, we must conclude that the body was never hot. Theory also suggests that perhaps the solar system was not formed in a hot state. Computers have calculated that the primordial nebula was relatively cool, with only the sun and the giant planets heating up when they formed, whereas the rest of the material remained cool.

▪ Some Open Questions

There are many outstanding questions. If the primordial nebula was cold, how can we explain the meteorites that show definite traces of earlier, albeit transient, heating?

Were the irregular, elliptical orbits of the asteroids—which often deviate greatly from circles—and their large inclinations caused by close encounters with the giant planets?

Why did no planet form between Mars and Jupiter? Is this why this region contains the asteroids, or in other words, kilometer-sized blocks of rock from which larger planets would otherwise have been formed?

How did the planets obtain their satellite systems? The outer, abnormal satellites are probably captured asteroids. But the inner, regular ones, such as the Galilean satellites of Jupiter, must have been formed by a process similar to the one that governed the formation of the planetary system itself.

According to the latest ideas, the sun must have begun as a rapidly rotating body. The difficulties with this are very similar to those found in Laplace's theory. In fact, there are many arguments supporting the theory that the sun once rotated faster around its axis. The T-Tauri stars, which we think resemble the young sun, rotate faster. Some kind of braking mechanisms probably slowed the sun down in the past, but we know very little about them.

In addition, we still have no explanation for the extreme inclinations of the axes of rotation of Uranus and Pluto. It is possible to imagine that collisions with other large bodies caused major changes in their rotation.

When we want to study these events in the remote past that led to the formation of the solar system, we are in a very difficult position, since we only know of one system: our own. We do not know if it is typical. It is possible that most cases are completely different, and that the formation of our system was a result of some extraordinary chance. Recently, we have obtained the first indications that other stars are probably still surrounded by primordial nebulae, from which planetary systems may be formed.

▪ *The Disk Around Beta Pictoris*

In the southern sky, we see the constellation of Pictor. Its second-brightest star, Beta Pictoris, has recently attracted attention. In Chapter 8, we mentioned one discovery made by the infrared satellite *IRAS*. We can now report another success for this American–European undertaking.

Bodies that are warm, but not so hot that they glow, mainly radiate in the infrared region, which is invisible to our eyes. *IRAS* was built specifically to detect radiation at wavelengths of a few thousandths of a millimeter. We can detect dust in the space between the stars, by such radiation. It does not glow and is heated only by the light from nearby stars. Neither does it radiate at radio wavelengths, and thus infrared radiation is the most suitable means of detecting it. *IRAS* detected masses of dust, of which we had only indirect indications previously. It also detected an excess of infrared radiation coming from the neighborhood of Beta Pictoris.

In 1984, two American astronomers, Bradford Smith of the University of Arizona and Richard Terrile of the Jet Propulsion Laboratory in Pasadena, succeeded in photographing a nebular disk around Beta Pictoris. Their attention was drawn to the star by the *IRAS* results, but they had to use a special technique in order to be able to determine from the surface of the earth, what caused the infrared excess. Beta Pictoris appears so bright that it prevents any weak features close to it from being seen. Therefore, the two astronomers blocked out its brilliant, telescopic image with a mask. They thus succeeded in photographing the immediate vicinity of the star without interference from its light. After tying many dodges, they obtained a photograph of the area around the star and saw two streaks of light, stretching out on both sides of the hidden image of the star. This was the appearance expected for the edge-on view of a disk surrounding a star. Like Saturn and its ring, Beta Pictoris is surrounded by a rotating disk. Its diameter corresponds to about 10 times the diameter of Pluto's orbit. Are we looking edge-on at a primordial nebula? Perhaps at this very moment, planets are being formed within it. Or have they perhaps already been formed?

The *IRAS* survey of the sky detected other stars suspected of having disks surrounding them that radiate in the infrared. Have planets already been formed within them? We shall soon know more.

Afterword

"No, the title of a book must directly indicate its subject," said Frau A., the editor of the German edition of this book. "With *Unheimliche Welten* [an approximate translation is 'Sinister Worlds'], the intent is not really clear. It might, for example, be some sort of horror story. Only the subtitle would indicate that it is an astronomical text."

We were sitting in the publisher's offices in Stuttgart. Dr. L., the chief editor, was not impressed with the title. "*Unheimliche Welten* gives the impression of danger, of something threatening, and that is certainly not the main accent of the book," he said.

"But isn't it dangerous when a planet has rain consisting of seething hot sulfuric acid?" I wanted to know. I persisted, not so much because I felt the title to be particularly suitable, but because other choices were even less to my liking. I was not very happy with my own title. Partly in order to convince myself, I continued:

"Don't be fooled by NASA's pictures that show the planets in delicate pastel shades, looking like some fairy-tale worlds. You cannot see from pictures of cloud-enshrouded Venus that the first breath of the air on that planet, which the ancients named after the goddess of love, would suffice to eat away one's lungs. An astronaut would probably last no longer on Mars if he were to leave the capsule without any protection. You have read my manuscript and know how inimical Jupiter and Saturn are to life, inimical because they were never created for man's occupation. Or, perhaps it would be better to say that he was never created to live on them. Those apparently innocuous points of light in the night sky are full of danger and threat for anyone that comes near them."

But Dr. L. also persisted. "The planets that you have written about also include the earth, the planet on which life has developed and on which we do, to a certain extent, feel safe. The earth is not 'sinister' for us." I gave in. For weeks I hunted for a new title. Dr. L. was right.

But then the thought occurred to me that the word "sinister" applied just as much to the earth as it did to any of the other planets. We may feel safe, but danger is still close at hand. When the earth began, everything was utterly inimical for life. The earth had no atmosphere when it condensed from the primordial nebula. The atmosphere was formed only later by gases from its interior. Water vapor and carbon dioxide collected on its surface. No air was retained by the earth's gravity from escaping into space that was fit to breathe, because there was a complete lack of oxygen for the first three billion years. But there were forms of life under what would be deadly conditions for us. Microorganisms that did not require oxygen began to release oxygen from carbon dioxide. The atmosphere slowly became enriched with oxygen. This gas was deadly to the early life forms. They produced what was a poison to them and exhaled it into the atmosphere. It would seem that all the oxygen now present in the atmosphere has been produced by life forms. About two billion years ago, this environmental pollution with oxygen caused a catastrophe. The atmosphere altered completely. But this drastic change took place so slowly that life had time to adapt. This was of the greatest value for its further development. In the new atmosphere, organisms arose that were able to derive their nourishment from the new environmental conditions. Multicellular and more and more complicated forms of life arose, eventually leading to modern man.

Life has itself created the atmosphere found on earth today. It governs the temperatures that reign on our planet. It is, in fact, an amazing situation. If the earth had a slightly smaller orbit around the sun, it would receive more radiation. It would be warmer, and more water would be evaporated from the oceans. The air would contain more water vapor, and infrared radiation would find it more difficult to escape into space. The earth would probably suffer from the greenhouse effect. However, if our planet had a slightly larger orbit around the sun, it would be colder. The white polar caps would enlarge and would reflect more sunlight back unused into space. The earth would become covered by ice. We should be glad that no one can alter the radius of the earth's orbit and can neither plunge us into a frozen polar waste nor turn the atmosphere into a greenhouse.

But life, which has changed the earth's atmosphere once, is well on the way to altering it once again. The carbon dioxide content of the atmosphere is rising. This time, however, since changes are not, as previously, taking place over billions of years but in mere centuries, life will not have time to adapt genetically.

If we continue to burn up fossil fuels, the atmosphere's carbon dioxide content will continue to rise. We do not know whether the oceans are capable of absorbing enough carbon dioxide to prevent us from heading toward conditions that resemble those prevailing on Venus.

Our increasing energy demands will cause even more oil to be burned, which only serves to release more carbon dioxide from the earth's crust into the atmosphere. The felling of forests also releases the carbon in plants into the atmosphere. Between the turn of the century and 1970, the carbon dioxide content of the atmosphere rose 7 percent. Since then, the rate of increase has accelerated. Carbon dioxide reduces the radiation of heat away from the earth. The temperature must increase, causing more water to evaporate from the oceans. Water vapor in the atmosphere increases cloud cover. Clouds reflect more sunlight than the earth's surface and therefore tend to counteract the warming. But they also restrict the earth's radiation into space, which in turn increases the greenhouse effect. We are uncertain which way the increased cloud cover will tip the balance. Is life on earth, which two billion years ago enriched the atmosphere in breathable oxygen, now turning the earth into a greenhouse?

It is rather surprising how many people, who protest about the impact of technology on nature, tend to concentrate on the fact that burning fossil fuels releases sulfur and lead into the environment, but relegate the carbon dioxide pollution of the atmosphere to second place. It may perhaps arise from a subconscious feeling that fossil fuels cannot be that dangerous, because even our grandparents burned coal or briquettes for heating. We are playing a dangerous game, because we really do not know how the atmosphere will be affected by man's pollution. In this scientific vacuum, ideology has free play. Thus, the question of using fossil fuels is closely related to the introduction of nuclear energy. Anyone who is against using fossil energy *and* against nuclear energy, must necessarily be in favor of using solar power, even though its proponents say that the technology is far from being sufficiently developed. "When things are uncertain, ideological dogma flourishes," a scientific expert wrote recently in a respected German weekly.

Previously, I felt that the use of nuclear energy was the only way of rescuing us from this dilemma. But as I wrote the first draft of the original German edition of this postscript, news came over the radio of the Chernobyl disaster.

If the other planets are "sinister worlds," then the earth also rightfully belongs in that category. Man has brought it to that state.

Appendix

A The Search for the Gravitational Constant

When a body moves in a curved path, such as a circle, it is subject to centrifugal force, which tends to push it outward. Any driver of a car feels this on a curve. I can feel it in my hand if I swing a weight on the end of a string rapidly around in a circle above my head.

If M_P is the mass of the body, R the radius of its circular path, and P the period that it takes for one circuit, then this force increases as the mass of the body and the radius of the orbit increase, but it decreases with increasing orbital period. Measuring the force on the string would show that it changes as

$$M_P \times R/P^2.$$

This is valid not only for a body whirled around on the end of a string, but also for a planet that is controlled by the sun's gravity. For simplicity, we will only consider planets that move in orbits closely approximating to a circle.

Kepler's third law, shown in Figure 12, is a relationship between the orbital diameter D and the orbital period P. The radius of the orbit R is half of D. Thus, the law can also be expressed as a relationship between R and P. It states that P^2 varies as R^3. Substituting this into the expression just given for the centrifugal force, we find that the latter varies as M_P/R^2.

If the centrifugal and gravitational forces are in balance (which they must be if the planet is not either to crash into the sun or to fly away from it), then the sun's gravitational attraction felt by the planet must be exactly the same as the centrifugal force. It must also vary as M_P/R^2 In the same way, the gravitational attraction must increase with increasing planetary mass M_P and decrease with increasing orbital radius. We already know that on earth, gravitational attraction increases with mass: Two sacks of potatoes weigh twice as much as one sack.

But the force between the sun and the planets is felt mutually. We should not say that the sun attracts a planet, but that the sun and the planet attract one another. With two bodies of initially equal mass, it is quite obvious that the force become twice as great if the mass of one of the two bodies is doubled. Applying this to the sun and a planet, it follows that the force must vary as the masses M_S and M_P of the sun and the planet, that is, as $M_P \times M_S/R^2$. This is not just valid for the sun and a planet but for any masses M_1 and M_2: Their mutual attraction varies as $M_1 \times M_2/R^2$.

In essence, this is nearly Newton's law of gravity, but not quite. As yet, I cannot calculate the attraction between two bodies. I only know how it varies if I vary the two individual masses and their distance. Expressing this in the language of school mathematics, the attraction k between two masses M_1 and M_2 at a distance R apart is proportional to $M_1 \times M_2/R^2$. Thus,

$$k = G \times M_1 \times M_2/R^2,$$

where G is a constant, a constant factor known as the gravitational constant. Once I know it, I can calculate the combined attractive force for any masses M_1 and M_2. If we are considering two extended, spherical bodies, then R is generally the distance between their centers. To calculate the force felt by a body at the surface of the earth, the earth's radius has to be substituted for R.

In order to establish the gravitational constant, the force between two bodies has to be actually measured. Let us take, for example, the attraction between the earth and a one-kilogram weight. This force amounts to 9.81 *Newtons*.[1]

Thus, we have $M_1 = 1$ kg, and $k = 9.81$ N, while R is equal to the earth's radius, that is, 6,370,000 meters. Newton did not know the mass of the earth, M_2, otherwise he could have obtained G from the equation given above.

Another possibility would be to take two masses, say two one-kilogram weights, to bring them together in the laboratory, and to measure their mutual attraction directly. Then, from the two masses, their distance, and the measured force, one could determine G from the above equation. Unfortunately, this force is not measurable: It is too weak. In fact, two one-kilogram weights at a distance of just ten centimeters experience a force that is less than one billionth of the force that each individual kilogram weight experiences acting toward the ground. In order to measure this tiny force, a special trick has to be used. The first person to think of this did not live to see his suggestion succeed.

The Englishman John Michell (1724–1793) made many significant contributions to astronomy. He was the first to estimate the distance of the fixed stars correctly and he recognized that they include double stars that are bound by their mutual gravitational attraction. He was also the first to realize that if a body had a sufficiently high density, no light could escape from it into space, because it would be forced to curve back onto it by the exceptionally high gravity. He is therefore the true father of black holes, which have become of such great interest in modern astrophysics, though the concept is generally ascribed to Laplace (see Chapter 7). Michell was professor of geology at Cambridge and later rector of a church near Leeds. William Herschel (see Chapter 4) may have learned to polish mirrors from him. Michell made the suggestion of how the weak force of gravitation might be measured.

Weak forces can only be measured when there are no strong forces opposing them. Experimentally, this can be arranged relatively easily. To understand what follows, imagine a mobile: A wire hanger is suspended horizontally by a thread, and from each end, another thread supports a further hanger, and so on, ending with brightly colored objects, all being finely balanced. The slightest movement of air sets the mobile in motion. The weak force of the draft affects the objects, and the horizontal hangers swing about their suspension points, twisting the threads. A thread strongly resists any pull tending to stretch it, but offers little resistance to a twisting force. John Michell had in mind the experiment shown schematically in Figure 82. A rod is suspended by a thread, with a weight at each end, carefully balanced so that it hangs perfectly level and with no tendency to rotate about its point of suspension. Now two heavy masses — spheres of lead, for example — are placed so close to the smaller, balanced masses that their attraction causes the rod to swing. The weights at the ends of the rod move closer to the lead spheres. Using a magnifying telescope, the motion of the ends of the rod can be observed against a scale behind them. The magnitude of the deviation

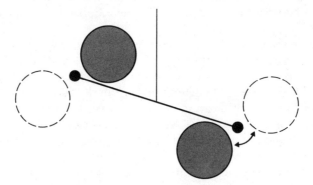

Figure 82. *The principle behind Cavendish's measurement of the gravitational constant. Each end of a suspended, balanced arm carries a small mass (small black spheres). Two lead spheres (grey) attract the masses carried by the arm. If the lead spheres are moved to different positions (dashed circles), the arm swings to a different direction. The amount of the swing allows the calculation of the strength of the attraction.*

indicates the force of attraction between the lead spheres and the smaller weights. The forces are so small that it takes minutes for any significant deviation to be detected.

The experiment was eventually carried out in 1798, after Michell's death, by the English physicist Henry Cavendish (1731–1810). When the heavy masses were brought close to the small weights, a weak attraction was detected. Cavendish was able to determine the gravitational constant in this way.

If, by using the equation given above, we express the masses in kilograms, the distance between the masses in meters, and the force in Newtons, then the value of G is a very small value, being the numeral 0, followed by the decimal point, and then ten zeros before the first numeral differing from zero. Although gravity is such a weak force, it plays a dominant part in the universe because of the enormous masses found in most celestial objects.

Once G is known, the mass of the earth can be determined. Let us assume that the two masses in the gravitational equation are a one-kilogram weight and the mass of the earth. We know the force exerted on a kilogram weight at the surface of the earth. We also know the earth's radius R and the gravitational constant G. Thus, we have all the values in Newton's gravitational equation except the mass of the earth, which we can now calculate. Therefore, ever since Cavendish determined the gravitational constant, we have known that we live on a body whose total mass, expressed in tons, is a 22-digit number.

Now the mass of the sun can be determined. The centrifugal force acting on the earth is known from the orbital radius and the orbital period; and using Newton's gravitational equation, we have all the values, except the mass of the sun. Since centrifugal force and gravity are equal, we can calculate the only unknown in the equation: the mass of the sun. This mass is equivalent to 300,000 times that of the earth.

B The Universal Clock

Anyone wanting to follow the evolution of stars and planets needs a clock that registers millions of years and billions of years, rather than hours, minutes, and seconds. Luckily, with no effort on our part, nature has provided us with just such a clock.

The principle is actually quite simple: Take a kilogram of the uranium isotope uranium 238 and leave it to its own devices.[1] If we look at it again after ten million years, a fraction of the uranium will have decayed into the lead isotope lead 206. Thus, we have just 995 grams of uranium and 5 grams of lead that has been formed from the uranium. After 4.5 billion years, half of the uranium will have decayed into lead. Therefore, if a celestial body contained just the isotope uranium 238 and no lead 206 when it was formed, then determining its lead content gives its actual age—a very simple procedure.

Unfortunately, detailed consideration reveals some snags. The method fails when the celestial body originally contained an unknown amount of lead as well as the uranium. We therefore find the lead that was originally present, plus the amount that has subsequently been created by the decay of uranium. Since these two amounts of lead only differ in their past history but not in their properties, we are unable to determine how much uranium has really decayed. Our method of determining the age will fail.

Our chronometer will also let us down even if no lead 206 was initially present. Uranium and lead differ in their physical and chemical properties: their atoms have different masses. So in a fluid mixture they will settle at different rates in a body's gravitational field. Because all planetary material was probably once fluid, there may be regions where the lead formed by decay happened to become concentrated, and other regions that are enriched in uranium. That would falsify any method of determining ages that depends on the relative abundances of uranium and lead.

Let us assume that lead was originally present. The moment the material containing the uranium and lead solidified, the two elements were unalterably locked together and could no longer be separated. If we knew their relative abundances when the material solidified, we would be able to determine the time that has elapsed from the current abundances. Over the course of time, the relative abundance would have shifted in favor of the lead.

But we do not know the initial relative abundances, because we cannot assume that conditions throughout the celestial body were exactly the same at the time the material solidified. And why should it have been exactly the same at the time it solidified? In the fluid state, the lighter lead would have floated on top of the heavier uranium, like droplets of fat on top of soup.

In order to get any further, we need to use another dodge. The lead isotope, lead 204, is not created by radioactive decay of other atoms. Since it hardly differs chemically and physically from lead 206, we can assume that even in the fluid state, the two types of lead always remained together. If any flow displaced one form of lead, it will also have displaced the other. Let us assume that we know the original relative abundances of the two lead isotopes, lead 206 and lead 204. Then the abundance of lead 204, which has remained unaltered since the material solidified, gives us the original abundance of lead 206. Any additional lead 206 that we find must have been created by radioactive decay of uranium. When we test samples with varying amounts of uranium 238 today, we can determine not only the original content of lead 206, but also—by using the current content of uranium 238—the time that has elapsed since the material solidified.

Unfortunately, we do not know the original relative abundances of the two lead isotopes. Nevertheless, we can still use this method of determining ages. If we want to determine the age of the earth, say, then we need to take several samples with differing contents of uranium 238 or lead 206. They must all have solidified at the same time, together with the materials in the earth's crust. If we investigate several samples (at least three), we do not just obtain sufficient equations to be able to determine the original relative abundances of lead 206 and lead 204, and thus the age, but we can also check whether all the samples solidified at the same time.

We have only discussed the decay of uranium 238 into lead 206. In fact, there are several radioactive clocks running simultaneously. The uranium isotope, uranium 235, turns into lead 207, and this is a clock with a relatively fast rate. In 700 million years, half of the uranium has turned into lead. The lead isotope, lead 208, is created from the element thorium. This clock has a very slow rate. Only after 14 billion years has half of the thorium decayed into lead.

A radioactive clock that runs about three times slower is one based on the rubidium isotope, rubidium 87, which changes into the strontium 87 isotope of strontium. This has been used to determine the age of six chondritic meteorites (see Chapter 12) as 4.54 billion years. Just as in the case of the uranium decay that we have been discussing, where the lead isotope, lead 204 (which is not created by radioactive decay) came to our assistance, here the strontium isotope, strontium 86, can be used.

C Radar and the Rotation of the Inner Planets

The technique of radar was developed just before World War II. During the war, it was used to locate enemy ships and aircraft by using aerials that emitted radio waves with wavelengths of a few decimeters.[1] These waves are reflected by many materials, especially metals. The reflected signal can be detected by a sensitive receiver using the same aerial. From the direction in which the signal was emitted and detected, the direction of the reflecting object is known. From the interval between the times of emission and detection of the signal the distance can be determined, because on both the outward and the return journey, the signal travels at the velocity of light. This means that it covers 300,000 kilometers every second. If the echo returns after one thousandth of a second, then the reflecting object is at a distance of 150 kilometers.

Using this method, ships at sea keep a watch for other vessels by means of continuously rotating aerials, and air-traffic controllers use radar to monitor aircraft in the vicinity of airports. Immediately after World War II, people succeeded in sending radar signals from the earth to the moon. The echo reached the earth after 2.5 seconds.

After radar techniques had been improved, people succeeded in obtaining echoes from Venus, Mercury, and the sun. The echo from the sun takes 16 minutes to return to the point on the earth from which it was emitted. This gave us a direct method of determining the distance between the earth and the sun. This method is quite unlike the one described in Chapter 3, which involves determining measurements of the solar system by using the parallax of minor planets. The radar distance confirmed the measurements made using Amor and Eros.

If a sharp radar pulse is sent toward Mercury, the return pulse is by no means sharp. The echo is broadened, that is, it is smeared out over a longer interval of time. The extent of the broadening can be used to determine the radius of the planet, because the time interval over which the returning pulse is received corresponds to the time that light requires to cover a distance twice the radius of the planet. For Mercury, this is slightly more than one hundredth of a second. Each individual point in time within the overall pulse corresponds to a single zone on the side of the planet turned toward us (Figure 83).

Each emitted radar pulse can be envisaged as a series of wavepeaks and troughs that move with the velocity of light. The number of peaks emitted in one second is the frequency of the signal. We might expect the echo to have the same frequency as the emitted signal, but this is only true if the planet is not moving relative to the earth.

Imagine that a signal emitted by us is returned by a reflector (Figure 84). As long as the reflector is stationary, the reflected peaks return at the same intervals as they were emitted.

But things are different when the reflector is moving toward the radar transmitter. It is then, so to speak, moving toward the wavepeaks. Each peak has a shorter distance to cover than its predecessor. The number arriving at the reflector per second has increased. What about the reflected peaks? Because the reflector is closer to the transmitter as it reflects each peak than it was for its predecessor, each wavepeak requires a shorter interval of time to reach the receiver, so the reflected peaks return at shorter intervals than they were emitted. It is easy to see that they will return at longer intervals if the reflector is moving away from the transmitter. This same principle of using radar to measure the velocity of a moving body is employed by the police to enforce speed limits on the highways.

Let us get back to the planet. We saw that at any particular instant within the overall echo, we received radiation that had been reflected by a particular zone (Figure 83). When the planet as a whole is moving toward or away from us, the frequency of the echo is shifted. If we transmitted a sharp pulse at a fixed frequency, then after a certain time, a pulse returns that is broadened with respect to time and that is also at a different frequency (Figure 85). If the planet is rotating, the situation is different again.

We again receive an echo that is spread out in time (Figure 86). At a time 1, we receive the echo from the tiny zone at the center of the side of the

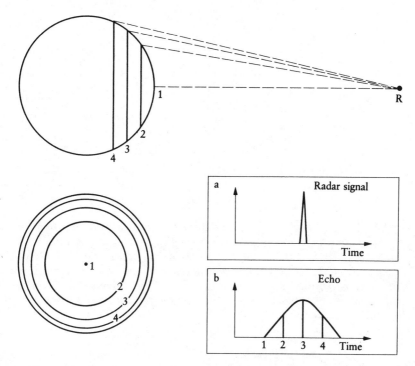

Figure 83. *A sharp radar signal (inset a) emitted toward a sphere is returned as a broadened echo (inset b). Top: The signal travels from the emitting aerial toward the sphere. It reaches different zones of the sphere's surface at different times. In the diagram, four zones 1, 2, 3, and 4 are shown. Their echoes return at different times, because the signals require different times for both the outward and return journeys. This is indicated in the broadened echo (inset b). Bottom left: The zones are rings. For the four zones shown, the echo from zone 1 returns first, followed by those from 2, 3, and 4 in order, and finally from the edge of the hemisphere that is turned toward the aerial.*

planet's disk facing us. Rotation is causing the motion of that particular spot on the planet's surface to be at right angles to the line of sight, but this has no additional effect on the frequency of the returning echo. This does not apply at time 2, however; we then receive the echo from the whole of zone 2, as shown in Figure 83. Because of the rotation, parts of this zone are moving toward us, and parts away from us. Thus, the frequency of the echo is correspondingly both increased and decreased. This means that the radar pulse that we sent out at a sharp, specific frequency returns broadened over a range of frequencies. This frequency broadening is greatest toward the end of the pulse (phase 4 in Figure 86). The alteration in the sharp, emitted pulse is dependent on the rotation and enables us to determine the speed of the

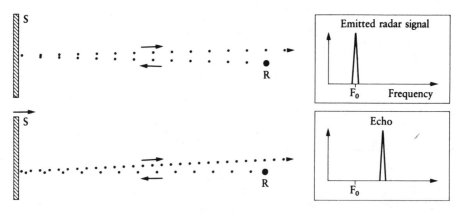

Figure 84. *The peaks of the waves in a train of pulses sent out at a frequency F_0 return at the same frequency (top) only if the reflector is stationary with respect to the aerial (top left). If it is moving toward the aerial, however (bottom left),* then the peaks return with a shorter interval between them, and the frequency is greater than that of the emitted signal *(bottom right). This is known as the* Doppler effect. *If the reflector is moving away, the frequency of the echo is lower than that of the emitted signal.*

planet's rotation. This was the method by which the rotation of Mercury and Venus was determined. It also enabled the direction of rotation to be detected.

We need to look at this last point in more detail. At first sight, it would appear impossible to tell whether a planet is rotating "toward the left" or "toward the right," because in either direction, the broadening would be the same, and it would not make the slightest difference whether the eastern edge or the western edge of the planetary disk is approaching us.

We begin with a simple example, considering a hypothetical inner planet that does not rotate as it revolves around the sun. For an observer on the planet, a fixed star would always remain at the same point on the sky. The planet's motion is shown schematically in Figure 87. If we observe the planet before, during, and after its inner conjunction, then it appears to rotate because we see it from different angles. It fools us into thinking that it rotates, even though it does not. We can call this its *apparent* rotation. It appears to occur in the same direction as its orbital motion. In Figure 87, anticlockwise orbital motion is shown. The nonexistent rotation is also anticlockwise. It is greatest at the time of inferior conjunction, because it is actually caused by the planet "moving past us," so that we see it from different directions. About a quarter of a revolution before and after this moment in time, no rotation occurs, because then the planet is moving

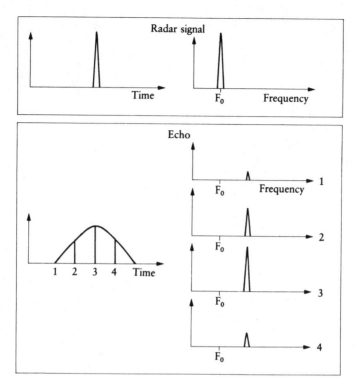

Figure 85. *If the sphere shown in Figure 83 is moving toward the observer, then the sharply defined pulse emitted at a frequency F_0 (top left) again produces a broadened echo (bottom left). In addition, the frequency of the echo is increased, because the sphere acts as a reflector that is moving toward the aerial (cf. Figure 84). The frequency of the echoes returned at the times 1, 2, 3, and 4 (bottom left) are shown (bottom right). All the echoes from individual ring zones are shifted an equal amount toward higher frequencies.*

toward or away from us, but not "past us." The apparent rotation is just as easily measured by radar as any true rotation.

If we investigate the radar echo from a planet, we measure both the true and the nonexistent rotation. We know, however, that the latter is greatest at inferior conjunction and that it occurs in the same direction as the orbital motion.

Let us assume that the actual rotation occurs in the same direction as the orbital motion. Because the measured rotational velocity is the sum of the true and the nonexistent rotations, we observe a rise in the measured velocity before conjunction. Afterwards, the measured velocity decreases; the nonex-

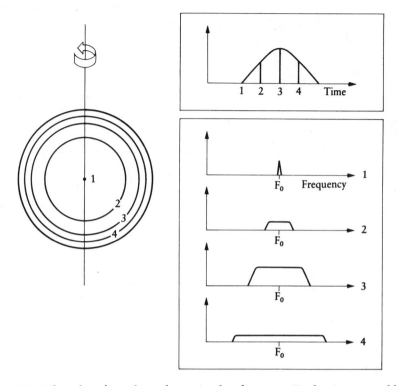

Figure 86. *The echo of a radar pulse emitted at frequency F_0 that is returned by a rotating sphere. (For simplicity, it is assumed that the sphere is moving as a whole either toward or away from the aerial.) The right-hand half of the disk is moving away from the aerial, and the left-hand half toward it. The four zones on the half of the sphere that is turned toward the aerial are indicated at left, and the broadened, overall echo is shown at top right. The times of the echoes returned by the center of the disk and the three zones are designated 1, 2, 3, and 4, as in Figure 83. The frequency distributions of the echoes returning at these four times are shown at bottom right. The echo at time 1 is returned by a zone that is neither moving toward the aerial, nor away from it. The frequency is the same as that of the emitted pulses. The echo at time 2 comes from a ring-shaped zone that because of the rotation is partly moving toward the aerial, and partly away from it. It therefore consists of moving "reflectors" (cf. Figure 84), and its frequency is partly increased and partly decreased. Thus, the signal is broadened in both time and frequency. The same applies to the echoes from zones 3 and 4. The strength of the frequency broadening enables the rotational velocity to be determined.*

istent rotation declines, whereas the true rotation remains the same. If, however, the orbital motion and the rotation are in opposite directions, then the nonexistent rotation tends to decrease the measured rotational velocity. Thus, the measured value declines before conjunction, reaches a minimum at conjunction, and increases afterwards. For Mercury, the measured rotation

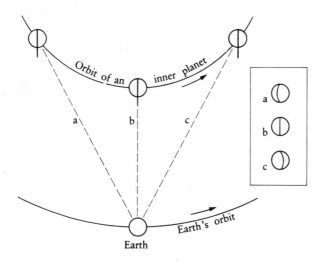

Figure 87. *How the direction of a planet's rotation can be determined from its radar echo. An inner planet may not rotate — that is, it may not rotate relative to the fixed stars. Then a meridian on the planet, here indicated together with a "flagpole," does not change its orientation relative to the stars. For an observer on earth, however, the planet would appear to rotate. At points a, b, and c on its orbit, the disk and the meridian of the planet would appear as shown in the inset. The meridian appears first to the left of the central line, and later on the right. The planet appears to be rotating in the same direction as its orbital motion. This apparent rotation can be observed by radar measurements. The way in which radar measurements may be used to determine the true rotation of a planet is explained in the text.*

reaches a maximum at (inferior) conjunction, whereas for Venus it reaches a minimum. Thus, Mercury rotates in the same direction as its orbital motion, but Venus does the opposite.

D   Radar Images of the Moon and the Planets

A telescope is not capable of achieving any desired degree of sharpness. More accurately, it is not always possible to separate two points on the sky that occur next to one another. Generally, through the telescope, the two points appear fused as a somewhat enlarged single spot. In that case, say that the telescope cannot *resolve* the two points and speak of a telescope's *resolving power.*

The resolving power of a telescope increases with increasing diameter of its objective. With a pair of binoculars, I could no longer resolve the pair of eyes in a man's face at a distance of some 6 to 10 kilometers, and the 5-meter (200-inch) reflector on Mt. Palomar would fail to do the same thing at a distance of over 600 kilometers.[1]

The resolving power of a radio telescope is far worse, because of the wavelength employed. With radio waves of 3-meter wavelength, which corresponds to VHF transmission, an aerial would have to have a diameter of 300 kilometers in order to achieve the same resolution as that of a pair of binoculars in visible light. Radar waves are shorter than VHF radio transmissions, but they are still half a million times longer than light waves. At first sight it would appear impossible to find out anything about the properties of the planets, say of Venus, from radar signals emitted from Earth.

In Appendix C, we saw that a short radar pulse sent out from earth toward a rotating planet is returned spread out in time. This happens because the signals take different times to reach the various zones on the planet and to return to us. We also saw that the frequency of the reflected radiation is altered. Now consider the side of the planet facing us (Figure 88). The planet may rotate with an equatorial velocity of 1 kilometer per second. Where are the points on its surface that are moving away from us at a velocity of, say, 200 meters per second? It is easy to see that these points lie on a chord across

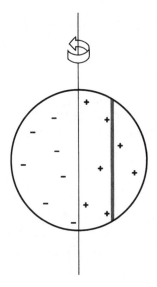

Figure 88. *Half of the points on the disk of a rotating planet are moving toward the radar aerial (− signs), and half are moving away (+ signs). All the points that are moving with the same velocity relative to the aerial lie on chords parallel to the axis of rotation. One of these is shown on the disk. Echoes of a pulse emitted by the aerial returning from all the points on such a chord have the same frequency shift.*

the disk, parallel to the axis of rotation. All the points on this chord are moving away from us with the same velocity. The reflected signals from all these points will be shifted to the same extent to lower frequencies (to longer wavelengths) by the Doppler effect. Thus, when we transmit a signal toward a planet, all the echoes that we receive with the same frequency shift come from a chord such as that shown in Figure 88.

Let us assume that on the surface of the planet there is a point (shown as A in Figure 89) that reflects radar waves particularly well in all directions. Let us also assume that all the other points absorb most radar radiation. The point therefore appears particularly bright at radar wavelengths. What does the echo of a sharp emitted pulse look like?

As the signal falls on the series of zones, the strength of the reflection will at first be low, until it reaches the zone shown shaded in Figure 89 (*left*). The point returns a strong echo. The strength of the echo then drops again for subsequent zones. Thus, we obtain an echo that appears like that shown in Figure 89 (*lower inset left*). The time at which the "spike" in the echo is

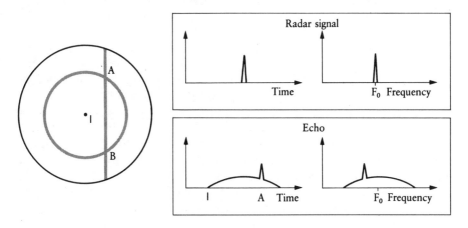

Figure 89. *A sharp radar pulse of frequency F_0 is beamed toward the surface of a rotating planet (top). Point A on the surface of the planet may be a particularly good radar reflector. Its echo returns after a specific interval. Before that, however, the echo returns from the closest point, 1, (bottom left), followed by the echoes from all the points that are closer than A. Finally, the echo returns from A. From the time of arrival of the signal, it is possible to determine that A lies on the zone corresponding to this interval of time. The shift in its frequency (bottom right) tells us from which chord on the planet's surface it originated. The strength of the echo at a particular time and at a specific frequency tells us the "radar brightness" of the point returning the echo. Unfortunately, the echo from point B has the same transit time and the same frequency shift. This is why the image of a planet's surface gained in this way is ambiguous.*

received tells us in which zone the bright spot occurs. But where does the spot actually lie on the zone?

In order to determine this we have to investigate the frequency distribution in the echo. We already know that if the planet was not moving the echo would return with precisely the same frequency. But if the planet rotates the frequencies of the echoes reflected by the various zones are significantly shifted by differing amounts. Since we assume that nearly all the points had very low reflectivity, a strong echo is returned by just our bright spot. Thus, from the frequency shift, we can determine which chord is returning the signal. This chord is shown in Figure 89 (*left*). The bright spot must lie where the ring zone and the chord intersect. This is how we can determine the location of bright spots on planetary disks from the earth.

We have only described the simplest case, where there is just one single bright spot on the surface of the planet. In reality, all the points return echoes of different strengths. By using the transit time and the frequency shift of every echo, it is possible to build up a picture of the individual reflecting points on a planetary disk.

The procedure has some flaws. We do not know whether the bright spot is really at point A as shown in Figure 89 (*left*). It could also be at point B, where the zone and chord also intersect. An echo from this point has the same time delay (point B is on the same ring zone), and exhibits the same frequency shift (point B is on the same chord). Thus, when we investigate the shift in time and frequency of an echo, we obtain a combined image of the northern and southern hemispheres of a planet, just as if we had a double exposure on a film.

The separation of the effects of the southern and northern hemispheres is particularly easy in the case of the moon. At first sight it might appear that the moon in not suitable for the use of radar techniques because it does not rotate. To be more accurate: It rotates so that it always turns the same face toward the earth during the course of its revolution around our planet. As the moon approaches the point in its orbit where it is nearest to us, all the points on its surface are moving toward us at the same velocity. We thus obtain the same frequency shift for every single point, and not, as with rotation, for just those that lie along a particular chord. But there are two effects that help us in the case of the moon. First, we are moving with the rotating earth, so during the course of a day, we see the moon from different directions. In the morning, we see slightly over its eastern edge, and in the evening slightly over the western edge. The daily rotation of the earth mimics a slight rotation of the moon. Second, we have already encountered the slight tumbling motion of the moon itself, known as libration (see Chapter 4). These two rotations (the daily one, arising from the earth's rotation, and the monthly one from libration) create frequency shifts in the radar echoes that allow us to determine which "chord" on the moon's surface is returning the radiation.

The techniques have been sufficiently developed so that it is possible to use the echoes of radar emission from earth to detect individual features on the surface of the moon that are 1 to 2 kilometers across. The best telescopes can do no better in visible light. In order to overcome the "double-exposure" effect, it is sufficient to direct the radar aerial once at the moon's northern hemisphere and once at the southern. The northern hemisphere appears brighter in one "double exposure" and the southern in the other, so the two can be distinguished.

For the planets, such as Venus, it is much more difficult to overcome the ambiguity. It requires a second radar receiver at a considerable distance from the first. If the two receivers are linked together electronically, then, in a certain sense, they act as a single large telescope. Linking radio telescopes did not only succeed in improving the resolving power of radio telescopes to such an extent that radio astronomers now "see" the sky in greater detail than their colleagues working in visible light; it also succeeded in solving the problem of the doubled images of Venus. The images of Venus obtained from earth were later fully confirmed by radar images from space probes.

Even from Earth, giant radar aerials were able to detect features on Venus that were of the order of a few kilometers in size. The radar images obtained from the *Venera 15* and *Venera 16* probes showed details down to about 1.2 kilometers across.

These two probes were inserted into long elliptical orbits around Venus in October 1983. For part of their 24-hour orbits, they came within 1000 kilometers of the surface of the planet, before swinging out as far as 65,000 kilometers from Venus. The probes' orbit is shown schematically in Figure 90, which shows that the orbit is at a right angle to the plane of the planet's equator. Each probe scanned a strip along the meridian when closest to the planet. When the probe was again 1000 kilometers from Venus, 24 hours later, the planet had rotated slightly, scanning a neighboring strip. Thus, the surface of Venus was gradually scanned, strip by strip, by each probe.

What happened each time they flew over the surface of the planet? Figure 91 shows the principle used for the measurements. As the probe flew along a meridian, 1000 kilometers above the surface, its aerial emitted radar waves, not directly beneath the probe's path but slightly to the right. The echoes of the various "spots" illuminated by the radar returned at various times. The first return came from point A, followed by echoes from circular arcs on the surface. Because the points on the arc were all at the same distance from the aerial, the travel times for the pulses were the same for each point. The last echo came from point B.

Just as with radar measurements from earth, the alteration in the frequency now had to be considered. The various points within the radar "footprint" were changing their distance from the aerial at different rates. The points on the straight line A–B were moving sideways relative to the probe. They were moving neither toward the aerial nor away from it, but at right angles to the

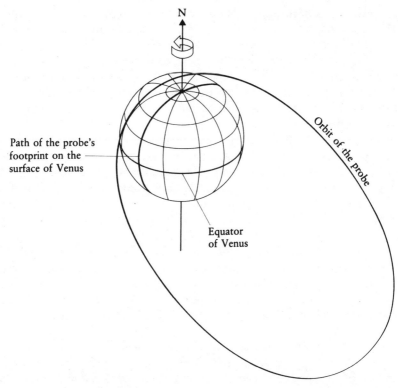

N

Path of the probe's
footprint on the
surface of Venus

Orbit of the probe

Equator
of Venus

Figure 90. *The orbits of* Venera 15 *and* 16. *When the probes were at the closest point in their orbits to the surface of Venus, at a height of 1000 kilometers, they surveyed a strip of the surface along a meridian with radar. When the probes returned after one orbit (24 hours later), Venus had rotated slightly, and the probes surveyed the neighboring strip.*

line of sight. The points in the forward half of the "footprint" were moving toward the probe, and the points in the backward half were moving away. This basically meant that all the points that had the same motion relative to the probe lay on essentially straight lines that intersected the arcs at right angles. The time at which an echo returned thus gave the arc on which a particular reflecting point lay, whereas the frequency shift gave the chord. The intersection of the two gave the position of the point. This is how the Soviet images (Figure 44) were obtained.

As already mentioned in Chapter 5, the American *Magellan* probe was launched in 1989, and apart from other objectives, it should provide us with new radar images from the surface hidden beneath the thick cloud cover around Venus.

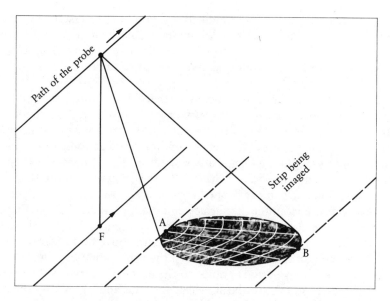

Figure 91. *The principles of the radar measurements made by* Venera 15 *and* 16. *As the probe flew over the surface of Venus at a height of 1000 kilometers, it "illuminated" an area to the right of its track with radar pulses. The interval before the receipt of an echo showed the distance from which it was being returned. Lines of equal distance from the aerial appear as circular arcs inside the radar "footprint." They correspond to the circular zones shown in Figure 86. The various points in the radar footprint are moving at different speeds relative to the aerial. The lines of constant velocity are not exactly straight lines but approximate them closely. They correspond to the chords shown in Figures 88 and 89. In a way precisely analogous to the procedure described in Figure 89, it is possible to determine the position of the origin of an echo from the interval and the frequency shift. The point on the surface of Venus immediately under the spacecraft is indicated by F. From the interval given by radar echoes from this point, the altitude profile of the area being crossed may be obtained.*

One of the greatest advantages of composing an image from the transit time and the frequency shift is that the method is not dependent on distance. With normal imaging methods (with photography, for example), the object appears smaller with increasing distance and can soon no longer be resolved by the camera. With radar images, we are only measuring the differences in the transit times and the frequency shifts. Both these values can be measured with a precision independent of the distance of the reflecting object. This is why the Venus probes did not show a very great increase in resolution over the radar images that had been obtained from Earth. Although the *Venera* probes were about 40,000 times closer to the planet's surface than radio telescopes on Earth, they only obtained a slightly better resolution. Where previously features down to a few kilometers could be detected, the probes were able to resolve objects of about one kilometer across. The *Magellan* measurements, however, should be able to resolve features down to a few hundred meters in size.

I still have to explain a phenomenon that Herr Meyer noticed while looking at the radar images of the surface of Venus (these can be seen in Figure 44 in Chapter 5). The radar images give a three-dimensional appearance to the various features. We get the impression that some of the mountain sides are in sunlight, and some are in shadow. This is because of the way in which the radar waves are reflected by the ground. A surface reflects incident radiation back to the probe best if it is at a right angle to the beam. If the radar waves encounter the surface at a shallow angle, they are reflected away in other directions. The echo received by the aerial is therefore very faint. Thus, when a mountain is illuminated from above by a radar beam, the side of the mountain that is turned toward the probe appears bright because the echoes returned by various points on the surface are strong. The echoes from the side opposite the probe are weak and thus appear dark on the radar image.

Notes

Chapter 1

[1] At first sight, the motion of the moon appears very regular, but in fact it is subject to many perturbations. As a result, the path of the moon varies over a period of years (see Chapter 4). We shall ignore this complication at this stage.

[2] The first edition of Copernicus' work appeared in about 500 copies. A second edition was published 20 years later. The astronomer and historian of science Owen Gingerich of Harvard University has tried to locate and examine all extant copies of the book. After exhaustive detective work, he found 260 copies of the first edition alone. He also discovered the copy once owned by Giordano Bruno, who was burnt at the stake in 1600 for ideas that had been prompted, in part, by this book, and copies that had belonged to Tycho Brahe and Johannes Kepler, both of whom we shall discuss later. Frequently, handwritten notes showed how thoroughly the owners had studied the book. The marginal notes in the owners' handwriting give some fundamental insights into how the individual hypotheses of Copernicus were received by the scholars in his day. Gingerich also found copies in the libraries of Henry II, king of France, and of Philip II, king of Spain.

Nearly half of the copies of the first edition are assumed to have been lost. Water and fire are the main enemies of old books. The loss of one particular copy is especially deplorable. A cargo of books was thrown overboard by Mediterranean pirates angry at such worthless booty. It included one of the Copernicus first editions. The current value of a single copy is some $50,000 to $60,000. Although most are owned by public libraries today, a few private collectors do acquire Copernicus' first editions. It is almost impossible to obtain a copy without Professor Gingerich knowing about it. Some time ago, he told me of a first edition owned by a couple in Frankfurt am Main. When I met him shortly afterwards, I learned that in the meantime, this copy had come onto the market and once more changed hands.

[3] The journalist and novelist Arthur Koestler might be right when he says in his biography of Johannes Kepler, *The Watershed*, that posterity probably glorifies Galileo more, simply because he had to appear before the Inquisition and to recant.

[4] Bertold Brecht wrote a very effective scene based on this idea into his play *The Life of Galileo*. The picture of the narrow-mined professors at the University of

Florence, however, seems rather too simple, too black and white, to be really convincing.

5 The revolution that Copernicus brought to our concept of the planetary system is one of the examples of how an opinion at odds with the commonly accepted view may triumph. Groucho Marx commented: "They claimed Galileo was mad when he said the earth orbited the sun, but he was right; they said the Wright brothers were crazy when they tried to fly, but they did; they said my uncle Wilbur was mad as a hatter—and he was!"

Chapter 2

1 Arthur Koestler has written "Not the least achievement of Newton was to spot the three laws in Kepler's writings, hidden away as they were, like forget-me-nots in a tropical flower bed."

Chapter 3

Epigraph The quotation from Jaroslav Hašek's *The Good Soldier Švejk* is from the translation by Cecil Parrott.

1 I often meet people who believe in astrology and who try to offer a rational explanation of why people who are born under a specific sign of the zodiac have similar characteristics. They say that the influences in the first months of life mold a person's character. A child born in winter will have a different character from one arriving in summer. The signs of the zodiac therefore merely indicate the date of birth during the year. Children born under the sign of Sagittarius are winter children, and those whose star sign is Aries arrived in spring. Because of the different seasons in the two hemispheres, those born under Aries in the southern hemisphere are autumn children and are therefore exposed to the same influences during their first months of life as children born in the northern hemisphere under Virgo. Yet astrologers apply their rules indiscriminately regardless of hemisphere of birth.

2 It is just as easy or as difficult to calculate the two values themselves from the ratio and the difference between the two as it is to solve the following problem: Jack is twice as old as his dog. Jack is seven years older than his dog. How old are Jack and his dog? Here again, we are dealing with the difference and the ratio between two values.

3 The Foucault pendulum is named after the French physicist Léon Foucault (1819–1868), who is 1851 hung a 28-kilogram weight at the end of a 67-meter-long steel wire from the dome of the Panthéon in Paris. The back-and-forth movement of the pendulum demonstrated the rotation of the earth. Before his

demonstration at the Panthéon, Foucault had tried the experiment with a line of only 2 meters and a weight of 5 kilograms.

Chapter 4

Epigraph The quotation from the *Chronica* of Gervase of Canterbury is translated by Dr. R. Hathorn.

[1] The reasons for eclipses are by no means generally understood, not even among so-called educated people. On one occasion, the geography teacher of one of my daughters had her students write the following in their exercise books: "From time to time, the moon passes through the earth's shadow; this causes the moon's phases." When I discovered this, I wrote underneath: "No, this is how eclipses of the moon are caused—the moon's phases occur for a different reason." When my daughter came back from school the next day, I asked her, "Did your teacher read it, and what did she say?" Answer: "She said 'Parents should not write in this notebook.'"

[2] Giovanni Domenico Cassini (1625–1712), professor of astronomy at Bologna and later director of the Paris Observatory.

[3] Johannes Hevel (1611–1687), also known as Helwelke or by his latinized name Hevelius, was an astronomer at Danzig.

[4] The English cleric Thomas Harriot (1560–1621) made a drawing of the moon through a telescope in July 1609, about six months before Galileo. He did not, however, realize that what he drew were mountain features.

[5] See Footnote 1 to Appendix B (below) for an explanation of the numbering scheme, using uranium as an example. The same applies to oxygen atoms. Its nucleus has eight protons and an additional number of neutrons. In oxygen 16, the nucleus has eight neutrons and eight protons. In oxygen 17, there is one additional neutron and so forth.

[6] Today, we can find the solar eclipse of 24 June 1778 in Oppolzer's *Canon* as number 7109.

Chapter 6

[1] Lowell's comment is from his book, *Mars and its Canals*, New York, 1906.

[2] M. Allaby and J. Lovelock, *The Greening of Mars*, André Deutsch, London, 1985.

Chapter 7

[1] Apianus's actual name was Peter Bienewitz; in accordance with the custom of the time, he had adopted a Latinized name.

[2] Ludwig Biermann was the first director of the Max Planck Institute for Astrophysics in Munich and therefore my predecessor in that post, to whom I am much indebted. In the last weeks of his life, he repeatedly told his friends how much he wanted to live to see *Giotto*'s encounter with Halley's comet. He pointed out that the event was due to occur on his birthday. Unfortunately, Ludwig Biermann, one of the greatest cometary researchers of our time, died eight weeks before his 79th birthday and the encounter with Halley.

[3] The "artificial comet" experiment was begun under the direction of Reimar Lüst, of the Max Planck Institute for Extraterrestrial Physics in Garching (Munich), and carried to a successful completion under the leadership of Gerhard Haerendel.

Chapter 8

[1] In astronomical reckoning, a century starts one year later than popularly assumed. This is because there was no year 0. The first century started with New Year's Day of the year 1. Astronomers will therefore begin the next millennium on the night of 31 December 2000 to 1 January 2001.

[2] Olbers was a physician and had introduced inoculation against smallpox in his hometown of Bremen, and had thus probably saved the lives of many Bremen citizens. He could well be described as the most successful amateur astronomer ever.

Chapter 9

[1] Because Jupiter orbits the sun once in approximately 11 years, which is the same period as the fluctuation in number of the dark areas on the sun, known as *sunspots*, laymen repeatedly assume that Jupiter must be responsible for these events on the sun. But the coincidence is just as accidental as the apparent agreement of the 28-day period of the moon's orbit around the earth with a woman's menstrual cycle, which caused astrologers to adopt the moon as the sign of fertility. Deducing some fundamental connection between cycles that have the same periods is about as sensible as assuming that there is some relationship between the four-year cycle once shown by maybugs (which have now, unfortunately, become rather rare) and elections to the German parliament,

which have precisely the same period. In any case, more accurate observations have shown that the solar cycle is not 11 years but 22.

[2] For the way nuclei are characterized, see Note 1 to Appendix B, below.

Chapter 11

[1] It was Colombo who suggested the name *Giotto* for the European Space Agency's (ESA) mission to Halley's Comet.

[2] When I recently saw a thick book about Pluto in a Munich bookstore, I picked it up immediately, because I wondered how it had been possible to find out so much about this distant planet. It turned out to be an astrological book. It explained the astrological significance of Pluto relative to all the other planets and the signs of the zodiac. Unfortunately, the author did not disclose how he had been able to find out so much, long before anyone had been able to observe Pluto's passage through all the signs of the zodiac. The author must have had a vivid imagination to be able to make it all up out of his own head.

The astrologer quoted at the beginning of this chapter also leaves us in the dark as to where she obtained her knowledge about Pluto. When questioned about where their rules come from, most astrologers answer that their knowledge dates back to antiquity, that the ancient Babylonians must have been highly sensitive to find out so much about the connections between men and the stars. Others say that in antiquity, people found out the influences of stars on men by scientific observations over hundreds of years. But things must be different for Pluto. All knowledge of its alleged influence must have been gained since 1930, not in the dim and distant past, but in the full light of recent decades. The advocates of astrology might like to explain how their knowledge of Pluto has been gained, especially those astrologers who like to give laymen the idea that they are scientists, by being photographed alongside a computer.

Chapter 12

[1] I have described this in more detail in my book *100 Billion Suns*.

Appendix A

[1] Although in everyday use, we commonly express forces in kilograms, to a physicist a *kilogram* (kg) is a measure of mass, and force is measured in *Newtons* (N). On earth, a mass of 1 kilogram experiences a downward force of 9.81 Newtons.

Appendix B

[1] There are various forms of uranium, known as *isotopes*. All atomic nuclei contain positively charged protons (their number being characteristic of each element) — 92 protons in the case of uranium — together with a varying number of electrically neutral neutrons (see also Footnote 5 to Chapter 4 above). In order to designate a particular uranium isotope, we give the name of the element followed by a number that gives the total number of particles in the atomic nucleus. Here we are dealing with a uranium isotope that has 92 protons and 146 neutrons, making a total of 238 particles.

Appendix C

[1] For radar waves, as for all forms of electromagnetic radiation, there is a simple relationship between wavelength and frequency. If the frequency is measured in millions of oscillations per second (the unit for this is the *megahertz*, abbreviated MHz), and the wavelength is measured in meters, then we have:

$$\text{Frequency in MHz} \times \text{wavelength in m} = 300$$

Thus for a wavelength of 30 cm (= 0.3 m), the frequency is 1000 MHz, or one billion oscillations per second. This equation also shows that radiation with shorter wavelengths has a higher frequency.

Appendix D

[1] In fact, it is more difficult to resolve objects with a telescope, because you have to look through the earth's atmosphere, and air movements cause the image to be unsharp.

Index